Solutions Manual

Integrated Mathematics for Explorers

Chee Leong Ching

Sun Jie

SriBooks

An Imprint of the Simplicity Research Institute, Singapore.
www.simplicitysg.net

Solutions Manual: Integrated Mathematics for Explorers.
Published by SRI Books,
an imprint of the Simplicity Research Institute, Singapore.
www.sribooks.simplicitysg.net

SriBooks

For bulk orders, special discounts, or to obtain customised versions of this book,
please contact SRI Books at **enquiry@simplicitysg.net**.

A CIP record for this book is available from the
National Library Board, Singapore.

Paperback Edition SRI-2015-2A.

ISBN: 978 − 981 − 09 − 5385 − 0 (pbook)
ISBN: 978 − 981 − 09 − 5386 − 7 (ebook)

Contents

Preface

This manual contains solutions (no questions) to selected questions from the book
Integrated Mathematics for Explorers by Adeline Ng and Rajesh R. Parwani:
Detailed solutions to all exercises.
Concise solutions to odd-numbered problems.
Answers to even-numbered problems are online at www.simplicitysg.net/books/imaths.

The material here is at a level suitable for high-school students in the GCE-O level or
IB programmes, or those in liberal arts colleges.

Topics covered include exponents, logarithms, polynomial equations, rational functions,
simultaneous equations, matrices, coordinate geometry, plane geometry, trigonometry, differential
and integral calculus.

Conventions: When units of length are not specified in a question, we usually do not insert
them into the solution. The reader should interpret the corresponding numbers for lengths, areas
and volumes accordingly.

About the Authors
Dr. Chee Leong Ching and Ms. Sun Jie are mathematics enthusiasts based in Singapore.

Acknowledgements
Chee Leong thanks Rajesh Parwani for the opportunity to work on this project.
Sun Jie dedicates this book to her mathematics teachers.

May 2015, Singapore.

Other titles by SRI Books and the Simplicity Research Institute:

Integrated Mathematics for Explorers by Adeline Ng and Rajesh R. Parwani
This book is for mathematics lovers in school, college and beyond. Topics are introduced through
real-life examples, accompanied by exercises at various levels of complexity. Challenges and
Investigations are suggested for the adventurous, while the Escapades chapter provides
stimulating puzzles, unsolved mathematical problems and beautiful theorems.

Real World Mathematics by Wei Khim Ng and Rajesh R. Parwani
This resource is for those who wish to learn or teach mathematics through real-world
applications. The problems are suitable for high schools and liberal arts colleges, with
accompanying notes summarising the background for the questions. Later chapters discuss the
scientific method and mathematical modelling.

School Mathematics (a series of volumes) by Chee Leong Ching, Sun Jie and Eksis Waiz
Contains brief review notes, worked examples, and test questions with answers. Questions are
selected or adapted from the two books mentioned above, with some new additions.

Simplicity in Complexity: An Introduction to Complex Systems by Rajesh R. Parwani
Topics covered include self-organisation, emergence, agent-based simulations, complex networks,
phase plane plots, fractals, chaos, and measures of complexity. Emphasis is placed on clarifying
common misconceptions.

Available from the publisher's webstore **www.store.simplicitysg.net** and other outlets.
Contact: **enquiry@simplicitysg.net**

Chapter 1

Exponents and Logarithms

1.1 Solutions to Exercises

1. *Note: Recall that any non-zero number may be written in scientific notation: $\pm A \times 10^p$ where $1 \leq A < 10$ and p is an integer.*
   We have,

   (a) $-0.0031 = -3.1 \times 10^{-3}$.

   (b) $123.456 = 1.23456 \times 10^2$.

   (c) $3.23 \times 12.56 \times 0.02 = 8.11376 \times 10^{-1}$.

   (d) $-24.31 \times 10^5 \times 0.07 \times 10^{-6}$
   $= -1.7017 \times 10^{-1}$.

---

2. *Note: Speed (v) is defined as the ratio of distance travelled to the time taken for the journey, $v = d/t$. Therefore $t = d/v$.*
   For the Moon-Earth case, the time taken is

   $$
   \begin{aligned}
   t_{\text{moon}} &= \frac{3.8 \times 10^5 \text{ km}}{3.0 \times 10^5 \text{ km/s}} \\
   &= \frac{3.8}{3} \\
   &= 1.3 \text{ s.}
   \end{aligned}
   $$

   By the same method,

   $$
   \begin{aligned}
   t_{\text{sun}} &= \frac{1.5 \times 10^8 \text{ km}}{3.0 \times 10^5 \text{ km/s}} \\
   &= \frac{1.5}{3.0} \times 10^{8-5} = 0.5 \times 10^3 \\
   &= 500 \text{ s} \\
   &\qquad \text{(since 1 min = 60 s)}. \\
   &= \frac{500}{60} \text{ min} \\
   &= 8.3 \text{ min.}
   \end{aligned}
   $$

   $$
   \begin{aligned}
   t_{\text{centauri}} &= \frac{4.0 \times 10^{13} \text{ km}}{3.0 \times 10^5 \text{ km/s}} \\
   &= \frac{4}{3} \times 10^{13-8} = \frac{4}{3} \times 10^8 \text{ s} \\
   &= \frac{\frac{4}{3} \times 10^8 \text{ s}}{365 \times 24 \times 60 \times 60 \text{s/yr}} \\
   &= 4.2 \text{ yrs.}
   \end{aligned}
   $$

   We used the conversions 1 yr = 365 days, 1 day = 24 hrs and 1 hr = 3600 s.

---

3. *Note: The decibel (dB) value is defined by $L = 10 \log_{10} \frac{I_1}{I_0}$ dB. Recall the identities $\log_{10} a = b \Rightarrow a = 10^b$ and $a \log_b c = \log_b c^a$.*

   (a) For $L = 20$dB,

   $$
   \begin{aligned}
   20 &= 10 \log_{10} \frac{I_1}{I_0} \\
   \Rightarrow \log_{10} \frac{I_1}{I_0} &= \frac{20}{10} = 2 \\
   \therefore \frac{I_1}{I_0} &= 10^2 \\
   &= 100.
   \end{aligned}
   $$

   Thus, the ratio of the two sound intensities is 100.

   (b) For $I_1 = 2I_0$,

   $$
   \begin{aligned}
   L &= 10 \log_{10} \frac{2I_0}{I_0} \text{ dB} \\
   &= 10 \log_{10} 2 \text{ dB} = \log_{10} 2^{10} \text{ dB} \\
   &= \log_{10} 1024 \text{ dB} \\
   &\approx 3.0 \text{ dB.}
   \end{aligned}
   $$

---

4. *Note: Recall that $\ln e^a = a \ln e = a \log_e e = a(1) = a$. Also, $\ln \frac{a}{b} = \ln a - \ln b$ and $\ln 1 = 0$.*

   (a) At the half-life point, the number of unchanged radioactive atoms is half of the initial

2

value, $N(t_{1/2}) = N_0/2$. Thus from the radioactive decay law, $N(t) = N_0 e^{-\lambda t}$,

$$N(t_{1/2}) = \frac{N_0}{2} = N_0 e^{-\lambda t_{1/2}}$$
$$\Rightarrow e^{-\lambda t_{1/2}} = 1/2.$$

By taking the natural logarithm of both sides, we obtain

$$\ln e^{-\lambda t_{1/2}} = \ln(1/2) = \ln 1 - \ln 2,$$
$$-\lambda t_{1/2} = 0 - \ln 2$$
$$\Rightarrow \lambda t_{1/2} = \ln 2$$
$$\therefore t_{1/2} = \frac{\ln 2}{\lambda}.$$

(b) Method I:
Consider the decay law $N(t) = N_0 e^{-\lambda t}$. We can obtain an expression for the time by taking the natural logarithm of both sides,

$$\ln N(t) = \ln(N_0 e^{-\lambda t}) = \ln N_0 + \ln e^{-\lambda t}$$
$$= \ln N_0 - \lambda t \ln e = \ln N_0 - (\lambda t)$$
$$\lambda t = \ln N_0 - \ln N(t) = \ln \frac{N_0}{N(t)}$$
$$\Rightarrow t = \lambda^{-1} \ln \frac{N_0}{N(t)}. \qquad (1.1)$$

Given that $N(t) = N_0/4$ hence $\ln \frac{N_0}{N(t)} = \ln 4$. Also, from part (a), we have $\lambda = \frac{\ln 2}{t_{1/2}} = \frac{\ln 2}{5730 \text{ yrs}}$. Substituting this into (1.1), we obtain

$$t = \frac{5730 \text{ yrs}}{\ln 2} \ln 4 = 2 \times 5730 \text{ yrs}$$
$$= 11,460 \text{ yrs}.$$

(We have used $\ln 4 = \ln 2^2 = 2 \ln 2$).

Method II:
When we look at the result of part (a), $t_{1/2} = \lambda^{-1} \ln 2$, it means that the half-life of a radioactive atom only depends on the type of element (which λ describes). It is not related to the number of radioactive atoms N_0 or $N(t)$. Therefore this question can be solved quickly without calculation. If $N(t) = N_0/4$, then it directly implies that we have gone through two half-lives: $N(t) = \frac{1}{2} \times \frac{1}{2} \times N_0$. Thus, the time involved should be $t = 2t_{1/2} = 2 \times 5730 \text{ yrs} = 11,460 \text{ yrs}$.

(c) From part (b) we can express the decay constant λ in term of $N(t), N_0$, and t as (1.1)

$$\lambda = t^{-1} \ln \frac{N_0}{N(t)}.$$

Given that $N(t) = 0.12 N_0$ and $t = 75,000$ yrs, the decay constant is

$$\lambda = \frac{\ln(1/0.12)}{75,000 \text{yrs}}$$
$$= \frac{\ln \frac{25}{3}}{75,000 \text{yrs}} = 2.83 \times 10^{-5}/\text{yrs}.$$

Then, from part (a), the half-life is

$$t_{1/2} = \frac{\ln 2}{\lambda} = \frac{\ln 2}{2.83 \times 10^{-5}/\text{yrs}}$$
$$= 2.45 \times 10^4 \text{ yrs}.$$

(d) The second radioisotope is plutonium-239.

---

5. (a) When a massive particle is stationary, $v = 0$, we have

$$E = \frac{mc^2}{\sqrt{1 - v^2/c^2}} \xrightarrow{v=0} mc^2,$$

which is known as the rest energy of the particle. (The speed of light in vacuum, $c = 3.0 \times 10^8$ m/s, is a fundamental constant of nature).

(b) If the energy E and mass m are real and finite (and $m > 0$), then the factor in the square-root must be positive, $(1 - v^2/c^2) > 0$. This implies that $v^2 < c^2$ or $0 \leq v < c$. Physically, it means that there is a maximum speed for a real massive particle and the upper bound of that maximum speed is given by $c = 3.0 \times 10^8$ m/s.

(c) [Optional]: As $v \to c$, the factor $\sqrt{1 - v^2/c^2}$ approaches zero and thus the energy of the particle diverges, $E \to \infty$. To obtain a physically meaningful finite energy as $v \to 0$ requires the particle to have zero mass, $m = 0$. In nature, there are such particles: For instance the photon, which is a quantum of light.

---

6. (a) For $n_1 = 1$ (Lyman series), Rydberg's formula is

$$\frac{1}{\lambda} = R\left(1 - \frac{1}{n_2^2}\right).$$

As λ increases, the left hand side decreases, and so must the right hand side. Therefore, to obtain the longest wavelength, we need to minimise the factor $(1 - 1/n_2^2)$. This is equivalent to maximising the factor $1/n_2^2$ (or minimising n_2). Since $n_2 > n_1 = 1$, we have to choose $n_2 = 2$. Note that both n_1, n_2 are positive integers. Hence, the longest wavelength is

$$
\begin{aligned}
\lambda_{\max} &= \left[R\left(1 - \frac{1}{n_2^2}\right) \right]^{-1} \\
&= \left[1.097 \times 10^7 \mathrm{m}^{-1} \left(1 - \frac{1}{2^2}\right) \right]^{-1} \\
&= \frac{4}{3} \times \frac{1}{1.097 \times 10^7} \mathrm{m} \\
&= 1.2 \times 10^{-7} \mathrm{m}.
\end{aligned}
$$

This is in the ultra-violet.

(b) For $n_1 = 2$ (Balmer series),

$$
\frac{1}{\lambda} = R\left(\frac{1}{4} - \frac{1}{n_2^2} \right).
$$

The shortest wavelength is radiated when the factor $(1/4 - 1/n_2^2)$ is maximised. This is equivalent to minimising the factor $1/n_2^2$, that is, $n_2 \to \infty$. Therefore the minimum wavelength for the Balmer series is

$$
\begin{aligned}
\lambda_{\min} &= \left[\frac{1.097 \times 10^7 \mathrm{m}^{-1}}{4} \right]^{-1} \\
&= 3.6 \times 10^{-7} \mathrm{m}.
\end{aligned}
$$

This is in the violet band.

(c) [Optional]: The human eye is sensitive to the visible band between 4×10^{-7} m to 7×10^{-7} m. So the human eye cannot detect the Lyman series but it might just be able to see the shortest wavelength of the Balmer's series.

---

7. (a) Since a^x is an increasing function of x for $a > 1$, therefore $7^5 < 7^7$, and similarly $5^{7/2} < 5^5$. Furthermore, $5^5 < 7^5$ since $1 < \left(\frac{7}{5}\right)^5$. Therefore we can arrange the terms in increasing order as

$$
5^{7/2} < 5^5 < 7^5 < 7^7.
$$

(b) Recall that $y = \log_a x$ is an increasing function of x for $a > 1$ and $x > 0$. In the question, all the logarithms have the same base $a = 2$ so we only need to compare the value of x. First, we rewrite the terms as $\log_2 3$, $\log_2 3^{1/2}$, $\log_2 3^2$, $\log_2 3^{-1}$. We have

$3^{-1} < 3^{1/2} < 3 < 3^2$. Therefore we can arrange the terms in increasing order as

$$
\log_2 \frac{1}{3} < \log_2 \sqrt{3} < \log_2 3 < \log_2 9.
$$

---

8. *Note: $(a^x)^y = a^{xy}$; and $a^x \times a^y = a^{x+y}$.*

(a) We first rewrite all the numbers in the same base,

$$
\begin{aligned}
8^{1/3} &= (2^3)^{1/3} = 2^1 . \\
4^{-3/2} &= (2^2)^{-3/2} = 2^{-3} . \\
\left(\frac{1}{2}\right)^{-2} &= (2^{-1})^{-2} = 2^2.
\end{aligned}
$$

Since $y = a^x$ is an increasing function of x if $a > 1$, we have $2^{-3} < 2^1 < 2^2$. Therefore

$$
4^{-3/2} < 8^{1/3} < \left(\frac{1}{2}\right)^{-2} .
$$

(b) As before, we rewrite all the numbers in same base,

$$
\begin{aligned}
(2^6)^{1/3} &= 2^2 . \\
(2^{1/2})^{-4} &= 2^{-2} . \\
2^2 \times 2^{-3} &= 2^{2-3} = 2^{-1} . \\
(2^2 \times 3^4)^{-1/2} &= (2^2)^{-1/2} \times (3^4)^{-1/2} \\
&= 2^{-1} \times 3^{-2}.
\end{aligned}
$$

We have $2^{-2} < 2^{-1} < 2^2$. Also, $2^{-1} \times 3^{-2} < 2^{-1} \times 2^{-1} = 2^{-2}$. Therefore,

$$
(2^2 \times 3^4)^{-1/2} < (2^{1/2})^{-4} < 2^2 \times 2^{-3} < (2^6)^{1/3}.
$$

---

9. (a) *Note: $(ab^x)^y = a^y b^{xy}$.*

$$
(2x^2)^{-1/2} (3x^{-5})^{-1}
$$
$$
= 2^{-1/2} \times x^{-1} \times 3^{-1} \times x^5 = \frac{1}{3\sqrt{2}} x^4.
$$

(b) *Note: $a \log_b x = \log_b x^a$ and $\log_a x + \log_a y - \log_a z = \log_a \frac{xy}{z}$.*

$$
\begin{aligned}
&\log_5 7 + 7 \log_5 \frac{1}{49} + \log_5 \sqrt{35} \\
&= \log_5 7 + \log_5 (7^{-2})^7 + \log_5 (7 \times 5)^{1/2} \\
&= \log_5 \left[7 \times (7^{-2})^7 \times (7 \times 5)^{1/2} \right] \\
&= \log_5 \left[7 \times 7^{-14} \times 7^{1/2} \times 5^{1/2} \right] \\
&= \log_5 \left[7^{-12.5} \times 5^{0.5} \right] \\
&= -12.5 \log_5 7 + 0.5 \log_5 5 \\
&= \frac{1}{2} - \frac{25}{2} \log_5 7
\end{aligned}
$$

4

where we made use of $\log_5 5 = 1$ in the last step.

(c) *Note:* $\frac{1}{\log_a b} = \log_b a$.

$$\log_3 25 + \frac{1}{\log_5 27}$$
$$= \log_3 5^2 + \frac{1}{\log_5 3^3}$$
$$= 2\log_3 5 + \frac{1}{3\log_5 3}$$
$$= 2\log_3 5 + \frac{1}{3} \times \frac{1}{\log_5 3}$$
$$= 2\log_3 5 + \frac{1}{3}\log_3 5$$
$$= \frac{7}{3}\log_3 5.$$

---

10. (a) *Note:* $\sqrt{a}\sqrt{b} = \sqrt{ab}$.

$$2\sqrt{27} + \sqrt{\frac{3}{4}} = \frac{2\sqrt{27}\sqrt{4} + \sqrt{3}}{\sqrt{4}}$$
$$= \frac{4\sqrt{3^2 \times 3} + \sqrt{3}}{2}$$
$$= \frac{4 \times 3 \times \sqrt{3} + \sqrt{3}}{2}$$
$$= \frac{13\sqrt{3}}{2}.$$

(b) *Note:* $(a - b)(a + b) = a^2 - b^2$.

$$\frac{5 - 2\sqrt{7}}{\sqrt{7} + 3} = \frac{5 - 2\sqrt{7}}{\sqrt{7} + 3} \times \frac{\sqrt{7} - 3}{\sqrt{7} - 3}$$
$$= \frac{11\sqrt{7} - 29}{7 - 9} = \frac{-29 + 11\sqrt{7}}{-2}$$
$$= \frac{29 - 11\sqrt{7}}{2}.$$

(c) *Note: For $a > 0$, $\sqrt{a^2} = a$ and $\frac{1}{\sqrt{a}} = \frac{\sqrt{a}}{a}$.*

$$5\sqrt{8} - \frac{1}{\sqrt{72}} + \sqrt{50}\sqrt{\frac{9}{2}}$$
$$= 5\sqrt{2^2 \times 2} - \frac{1}{\sqrt{6^2 \times 2}} + \sqrt{5^2 \times 2}\sqrt{\frac{3^2}{2}}$$
$$= 10\sqrt{2} - \frac{1}{6\sqrt{2}} + 5\sqrt{2} \times \frac{3}{\sqrt{2}}$$
$$= 10\sqrt{2} - \frac{\sqrt{2}}{12} + 15$$
$$= \frac{119}{12}\sqrt{2} + 15.$$

11. (a) *Note:* $a^x = a^y \Rightarrow x = y$.

$$5^{2x-1} = 1/25 = \frac{1}{5^2} = 5^{-2}$$
$$\Rightarrow 2x - 1 = -2$$
$$\therefore x = -\frac{1}{2}.$$

(b)

$$3^{2-x}(2^2 \times 3^{2x+1}) = 4/9 = 2^2 \times 3^{-2}$$
$$3^{2-x} \times 3^{2x+1} = 3^{-2}$$
$$3^{(2-x)+(2x+1)} = 3^{-2}$$
$$3^{x+3} = 3^{-2}$$
$$\Rightarrow x + 3 = -2$$
$$\therefore x = -5.$$

(c) *Note:* $\log_a x = b \Rightarrow x = a^b$.

$$\log_3(2x - 1) = 2$$
$$2x - 1 = 3^2$$
$$\therefore x = 10/2$$
$$= 5.$$

(d)

$$\log_{10}(x + 1) - \log_{10}(x - 1) = 2 - \log_{10} 2$$
$$\log_{10}\frac{(x + 1)}{(x - 1)} + \log_{10} 2 = 2$$
$$\log_{10}\frac{2(x + 1)}{(x - 1)} = 2$$
$$\frac{2(x + 1)}{(x - 1)} = 10^2 = 100$$
$$\Rightarrow x + 1 = \frac{100}{2}(x - 1)$$
$$= 50(x - 1)$$
$$\Rightarrow 49x = 51$$
$$\therefore x = \frac{51}{49}.$$

(e) *Note:* $\log_b x = 1/(\log_x b)$.

$$\frac{\log_2 x^2}{2\log_x 2} = 9$$
$$\frac{2\log_2 x}{2\log_x 2} = 3^2$$
$$\log_2 x = 3^2\log_x 2$$
$$= \frac{3^2}{\log_2 x}$$
$$\Rightarrow (\log_2 x)^2 = (3)^2$$
$$\log_2 x = \pm 3$$
$$\therefore x = 2^{\pm 3}$$
$$= 8 \text{ or } \frac{1}{8}.$$

(f)

$$\begin{aligned}
\sqrt{x-1} &= (x-3) \\
\Rightarrow x-1 &= (x-3)^2 \\
x-1 &= x^2 - 6x + 9 \\
0 &= x^2 - 7x + 10 \\
&= (x-2)(x-5) \\
x &= 2 \text{ or } 5.
\end{aligned}$$

However, since we have squared the original equation, we need to verify if both solutions are acceptable (see page 9 of the book). Substitution into the original equations shows that only $x = 5$ is a solution.

(g)

$$\begin{aligned}
7^3 &= e^x \\
\ln(7^3) &= \ln(e^x) \\
\therefore x &= 3\ln 7.
\end{aligned}$$

12. (a) Recall the Binomial Theorem,

$$(a+b)^n = \sum_{r=0}^{n} \binom{n}{r} a^{n-r} b^r$$

where $\binom{n}{r} \equiv \frac{n!}{(n-r)!r!}$. By setting $a = 1; b = 0.02$, we have

$$\begin{aligned}
(1.02)^4 &= 1^4 + \binom{4}{1} 1^3 \times (0.02)^1 \\
&\quad + \binom{4}{2} 1^2 \times (0.02)^2 \\
&\quad + \binom{4}{3} 1^1 \times (0.02)^3 + ... \\
&\approx 1 + 4 \times 0.02 = 1.08 \\
&\approx 1.1 \ .
\end{aligned}$$

(b) As in part (a), we rewrite the term as $(2.02)^6 = (2.00 + 0.02)^6 = 2^6(1.00 + 0.01)^6$. We have,

$$\begin{aligned}
(2.02)^6 &= 2^6 \left[1^6 + \binom{6}{1} 1^5 \times (0.01)^1 + ... \right. \\
&\quad \left. + \binom{6}{5} 1^1 \times (0.01)^5 + (0.01)^6 \right] \\
&\approx 2^6 \left[1 + 6 \times (0.01)^1 + 15 \times (0.01)^2 \right] \\
&= 2^6 \left[1 + 0.06 + 0.0015 \right] = 67.936 \\
&\approx 67.9
\end{aligned}$$

where in the expansion we kept only the leading order terms.

13. (a)

$$\begin{aligned}
&\left(1 + \frac{x}{2} \right)^6 \\
&= 1^6 + \binom{6}{1} 1^5 \times \left(\frac{x}{2} \right)^1 \\
&\quad + \binom{6}{2} 1^4 \times \left(\frac{x}{2} \right)^2 + ... \\
&= 1 + 3x + \frac{15}{4} x^2 + ...
\end{aligned}$$

Thus, the expansion to the first three terms in increasing powers of x is $1 + 3x + \frac{15}{4} x^2$.

(b)

$$\begin{aligned}
&(1 + 2x^2)^9 \\
&= 1^9 + \binom{9}{1} 1^8 \times (2x^2)^1 \\
&\quad + \binom{9}{2} 1^7 \times (2x^2)^2 + ... \\
&= 1 + 9 \times (2x^2) + 36 \times (4x^4) + ... \\
&= 1 + 18x^2 + 144x^4 + ...
\end{aligned}$$

(c)

$$\begin{aligned}
\left(x + \frac{1}{x} \right)^8 &= \frac{1}{x^8} (1 + x^2)^8 \\
&= \frac{1}{x^8} \left[1 + \binom{8}{1} (x^2)^1 + \binom{8}{2} (x^2)^2 + ... \right] \\
&= \frac{1}{x^8} (1 + 8x^2 + 28x^4 + ...) \\
&= \frac{1}{x^8} + \frac{8}{x^6} + \frac{28}{x^4} + ...
\end{aligned}$$

(d)

$$\begin{aligned}
&(2 + x)^6 (1 + x)^2 \\
&= 2^6 (1 + 2x + x^2) \left(1 + \frac{x}{2} \right)^6 \\
&= 2^6 (1 + 2x + x^2) \left[1 + \binom{6}{1} \left[\frac{x}{2} \right]^1 \right. \\
&\qquad \left. + \binom{6}{2} \left[\frac{x}{2} \right]^2 + ... \right] \\
&= 2^6 (1 + 2x + x^2) \left[1 + 3x + \frac{15}{4} x^2 + ... \right] \\
&= 2^6 \left[\left(1 + 3x + \frac{15}{4} x^2 \right) + 2x(1 + 3x + ...) \right. \\
&\qquad \left. + x^2(1 + ...) + ... \right] \\
&= 64 + 320x + 688x^2 + ...
\end{aligned}$$

(e)

$$x(1+3x)\left(1+\frac{1}{x}\right)^6 = x(1+3x)\frac{1}{x^6}\left(1+x\right)^6$$

$$= \frac{1}{x^5}(1+3x) \times \left[1 + \binom{6}{1}x + \binom{6}{2}x^2 + ...\right]$$

$$= \frac{1}{x^5}(1+3x) \times (1 + 6x + 15x^2 + ...)$$

$$= \frac{\left[(1 + 6x + 15x^2 + ...) + 3x(1 + 6x + ...)\right]}{x^5}$$

$$= \frac{1}{x^5}(1 + 9x + 33x^2 + ...)$$

$$= \frac{1}{x^5} + \frac{9}{x^4} + \frac{33}{x^3} + ...$$

---

14. (a) From Fig.(1.1), we see that there is only one solution to the equation $3\ln x + 2x - 1 = 0$. (Note that $3\ln x$ is an increasing function.)

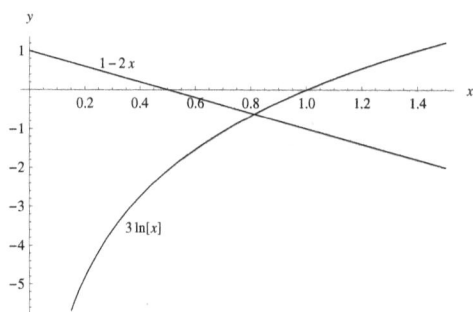

Figure 1.1: Plot of both $f(x)$ and $g(x)$.

(b) An accurate plot suggests that the numerical solution is $x \approx 0.81$.

---

15. (a) Fig.(1.2) shows that there are two solutions to the equation $2^x = 8x(1-x)$ or equivalently $2^{x-3} = x(1-x)$ in the range $0 \leq x \leq 1$.

(b) An accurate plot suggests that the numerical solution in the range of $0 \leq x \leq 1/2$ is $x \approx 0.17$.

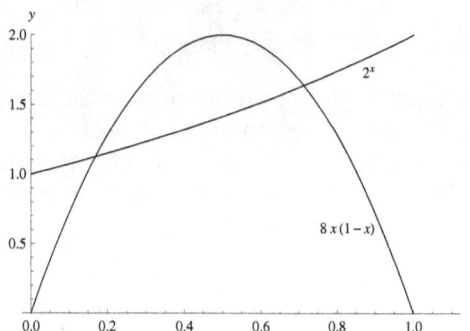

Figure 1.2: Plot of both $f(x)$ and $g(x)$.

Did You Know?

The constants e and π are linked by the "imaginary" number $i = \sqrt{-1}$ through the amazing relation $e^{i\pi} = -1$.

Formula List

You can download brief review notes on school mathematics, including a formula list, for **free** from www.simplicitysg.net/books/imaths.

1.2 Solutions to Odd-Numbered Problems

1. (a) let $N(t)$ be the number of transistors in year t. We have $N(t+2) = 2N(t)$. So, $N(2) = 2N(0)$, $N(4) = 2^2 N(0)$ and so on. This trend can be summarised by

$$N(t) = 2^{\frac{t}{2}} N_0.$$

where we have denoted $N(0)$ by N_0. To increase by 10-fold, we have $N(t) = 10N_0$, so

$$
\begin{aligned}
10N_0 &= 2^{\frac{t}{2}} N_0 \\
\log 10 &= \frac{t}{2} \log 2 \\
\therefore t &= 2 \times \frac{\log 10}{\log 2} \\
&= 6.64 \text{ yrs.}
\end{aligned}
$$

---

3. (a) Similar to problem (1), we have $N(t) = 2^{\frac{t}{12}} N_0 = 2^{\frac{t}{12}} \times 5$, where t is measured in minutes. After an hour, $t = 60$ min, the number of cells is $N = 2^{\frac{60}{12}} \times 5 = 160$.

(b) To reach $N = 2000$,

$$
\begin{aligned}
2000 &= 2^{\frac{t}{12}} \times 5 \\
2^{\frac{t}{12}} &= 400 \\
t &= \frac{12 \log 400}{\log 2} \\
&= 103.7 \text{ min.}
\end{aligned}
$$

---

5. (a) Recall that $\log_2 x$ is an increasing function of x. Since $\sqrt{3} < 3 < 6 < 9$, we get $\log_2 \sqrt{3} < \log_2 3 < \log_2 6 < \log_2 9$.

(b) First, rewrite the terms in base 2. We have $\log_4 3 = \frac{\log_2 3}{\log_2 4} = \frac{\log_2 3}{\log_2 2^2} = \frac{\log_2 3}{2} = \log_2 \sqrt{3}$. Also, $2\log_9 3 = \frac{2\log_2 3}{\log_2 9} = \frac{2\log_2 3}{\log_2 3^2} = 1 = \log_2 2$. Since $\sqrt{3} < 2 < 2\sqrt{3}$, we have $\log_4 3 = \log_2 \sqrt{3} < 2\log_9 3 < \log_2 2\sqrt{3}$.

---

7. (a) Let $y = e^x$. The equation becomes

$$
\begin{aligned}
0 &= y^2 - (e^3 + 2)y + 5 \\
y &= \frac{(e^3 + 2) \pm \sqrt{(e^3 + 2)^2 - 20}}{2} \\
&= 0.229 \text{ or } 21.86 \\
\therefore x &= \ln y \\
&= -1.48 \text{ or } 3.08.
\end{aligned}
$$

(b) Let $y = 2^x$ and rewrite the equation as

$$
\begin{aligned}
3(y^2 - 1) &= \frac{y}{2} - \frac{3y^2}{16} \\
\Rightarrow 51y^2 - 8y - 48 &= 0 \\
y &= \frac{4 \pm 4\sqrt{154}}{51}.
\end{aligned}
$$

Since $x = \ln y / \ln 2$, we have $x = \dfrac{\ln \frac{4+4\sqrt{154}}{51}}{\ln 2} = 0.07$. (As $y = 2^x > 0$ for all x, the case $y = \frac{4-4\sqrt{154}}{51}$ is not a valid solution).

(c) Let $y = \sqrt{5^{-x}}$ and rewrite the equation as

$$
\begin{aligned}
2 &= 3 \times 5y^2 - 9 \times \frac{1}{5y} \\
\Rightarrow 2 &= 15y^2 - \frac{9}{5y}.
\end{aligned}
$$

We can solve the equation numerically by the graphical method, plotting $f = 2 + 9/(5y)$ and $f = 15y^2$ on the same graph and seeking the intersection points. See the figure below.

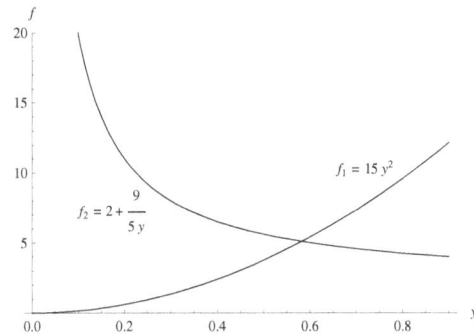

Figure 1.3: For problem 7c.

From the figure, we obtain $y \approx 0.5825$. Hence $x = \frac{-\ln y^2}{\ln 5} = 0.67$.

(d) Let $y = \lg x$. Thus we have

$$
\begin{aligned}
0 &= (\lg x)^2 + \lg \sqrt{x} - 0.1 \\
&= y^2 + \frac{y}{2} - 0.1 \\
\Rightarrow y &= \frac{-0.5 \pm \sqrt{0.65}}{2} \\
&= 0.153 \text{ or } -0.653 \\
\therefore x &= 10^y \\
&= 1.42 \text{ or } 0.22.
\end{aligned}
$$

(e) Let $y = 2^x$, we have

$$2x = \log_2(2^{x-1} - 3 \cdot 4^x)$$
$$2^{2x} = \frac{2^x}{2} - 3 \cdot 2^{2x}$$
$$y^2 = \frac{y}{2} - 3y^2$$
$$\Rightarrow y = 0 \text{ or } \frac{1}{8}$$

Since $y = 2^x > 0$ for finite x, only $y = 1/8$ is valid. Therefore $x = \log_2 2^{-3} = -3$.

(f)

$$\ln\left|\frac{\sqrt{x} + \sqrt{3}}{\sqrt{x} - \sqrt{3}}\right| = -\frac{1}{2}\ln\frac{1}{|x|} = \ln\sqrt{|x|}$$
$$\left|\frac{\sqrt{x} + \sqrt{3}}{\sqrt{x} - \sqrt{3}}\right| = \sqrt{|x|} .$$

Note the constraint $x \geq 0$ since we have $\sqrt{x}$ in the expressions. So we can ignore the modulus on right hand side. We can drop the modulus on left hand side by considering the two possibilities

$$\pm\left(\frac{\sqrt{x} + \sqrt{3}}{\sqrt{x} - \sqrt{3}}\right) = \sqrt{x}.$$

Let us consider the " $+$ " case first. Denote $y = \sqrt{x}$, we have

$$\frac{y + \sqrt{3}}{y - \sqrt{3}} = y$$
$$0 = y^2 - (\sqrt{3} + 1)y - \sqrt{3}$$
$$\Rightarrow y = \frac{(\sqrt{3} + 1) \pm \sqrt{4 + 6\sqrt{3}}}{2}$$
$$= 3.263 .$$
$$\therefore x = y^2$$
$$= 10.65 .$$

(We ignored the second solution $y = -0.501$ since $y = \sqrt{x} \geq 0$).

You can check that the " $-$ " case does not give valid solutions.

(g)

$$\lg\sqrt{x^2 - 3x + 2} = 1/\log_x 10 = \lg x$$
$$\sqrt{x^2 - 3x + 2} = x$$
$$x^2 - 3x + 2 = x^2$$
$$x = \frac{2}{3}.$$

(h)

$$0 = \log_x \log_x \log_x 2^{3x}$$
$$x^0 = 1 = \log_x \log_x 2^{3x}$$
$$x^1 = x = \log_x 2^{3x} = 3x \log_x 2$$
$$\Rightarrow \log_x 2 = 3^{-1}$$
$$2 = x^{1/3}$$
$$\therefore x = 8.$$

---

9. Rewrite both equations. First,

$$27^x \cdot 3^{5y} = 9$$
$$3^{3x} \cdot 3^{5y} = 3^2$$
$$3^{3x+5y} = 3^2$$
$$\Rightarrow 3x + 5y = 2 . \qquad (1.2)$$

For the second equation,

$$\log_2 y = \log_2(1 - x) - 1$$
$$\log_2 \frac{y}{1 - x} = -1$$
$$\Rightarrow y = \frac{1 - x}{2}. \qquad (1.3)$$

The two simultaneous equations (1.2) and (1.3) can be solved to give $x = -1; y = 1$.

---

11. (a) Square both sides to get

$$8 + 2\sqrt{16 - x^2} = 16 - x^2 .$$

Now let $y = \sqrt{16 - x^2} \geq 0$, then

$$0 = y^2 - 2y - 8$$
$$\therefore y = 4.$$

Since $y = \sqrt{16 - x^2}$, this implies that $x = 0$.

(b) Square both sides to get

$$\sqrt{x^2 - 5x + 4} = 10 - 2x$$
$$\Rightarrow x^2 - 5x + 4 = 100 - 40x + 4x^2$$
$$x = \frac{35 \pm \sqrt{73}}{6}.$$

However, direct substitution into the initial equality shows that only $x = \frac{35 - \sqrt{73}}{6}$ is valid.

(c) Re-arrange the equation and take the square of both sides,

$$2x^2 - x - 3 = x^2 + 2$$
$$x = \frac{1 \pm \sqrt{21}}{2}.$$

(d) First, factorise the expressions to get

$$\sqrt{(x-5)(x+1)} + \sqrt{(x-3)(x+1)} = 2(x+1)$$

Obviously $x = -1$ is a solution. If $x \neq -1$, divide both sides by $\sqrt{x+1}$ and solve the reduced equation by the methods of previous problems. However no new valid solutions are found.

---

13. Let the radius be $r = a + b\sqrt{3}$ where a, b are rational numbers. Therefore

$$4 + 2\sqrt{3} = (a^2 + 3b^2) + 2ab\sqrt{3}.$$

This gives two simultaneous equations that we can solve for a and b to get $r = 1 + \sqrt{3}$.

---

15. (a) We have $P(n) = (1.03)^n P_0$ where P_0 is the initial amount deposited and n is the number of months the money is kept. For $n = 3$ we have

$$
\begin{aligned}
P &= (1 + 0.03)^3 (10^4) \\
&= \left[1 + 3(0.03) + 3(0.03)^2 + (0.03)^3\right] (10^4) \\
&= 10,927.27.
\end{aligned}
$$

(b) We get the same answer using a calculator.

---

17. (a) Using the binomial theorem,

$$\left(1 + \frac{x}{2}\right)^{10} = 1^{10} + \binom{10}{1}(1)^9 (x/2)^1 + \dots$$

so the term independent of x is $1^{10} = 1$.

(b) As in part (a),

$$\left(1 + (2x)^2\right)^{11} = 1^{11} + \binom{11}{1}(1)^{10}(2x^2)^1 + \dots$$

so the term independent of x is $1^{11} = 1$.

(c) The binomial theorem can be written as

$$(a+b)^n = \sum_{r=0}^{n} \binom{n}{r} a^{n-r} b^r$$

For $(x + 1/x)^8$, the term independent of x arises when $8 - r = r$, so $r = 4$. Explicitly, it is $\binom{8}{4} x^4 (1/x)^4 = 70$.

(d) Using the binomial theorem

$$
\begin{aligned}
(1+3x)^2(2+x)^6 &= 2^6(1+3x)^2(1+x/2)^6 \\
&= 2^6(1 + 6x + 9x^2)(1 + 3x + \dots),
\end{aligned}
$$

so the term independent of x is $2^6 = 64$.

19. We can solve the equation numerically by plotting both $f_1(x) = e^{-x}$ and $f_2(x) = \sin x$ on the same graph and looking for the intersection points.

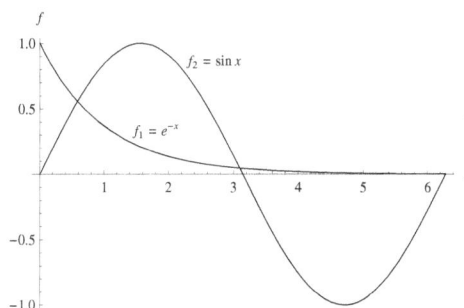

Figure 1.4: For problem 19.

The sketch shows that there are exactly two solutions between $0 \le x \le \pi$ and no solution between $\pi \le x \le 2\pi$. From the figure, the solution in the range $0 \le x \le \pi/2$ is given by $x \approx 0.59$.

Did You Know?

The E-book edition of *School Mathematics: Volume 1* is available for **free**!

The volumes, produced by SRI Books, contain selected questions with detailed solutions, and test questions with answers.

Visit the publisher's website www.sribooks.simplicitysg.net to find out more.

Chapter 2

Polynomials and Rational Functions

2.1 Solutions to Exercises

1. *Note:* $ax^2 + bx + c = 0 \Rightarrow x = \frac{-b \pm \sqrt{b^2 - 4ac}}{2a}$.

   (a) Let $\frac{a}{b} = x$, we can then rewrite the equation in the question as

   $$\frac{a}{b} = \frac{a+b}{a} = 1 + \frac{a}{b}$$
   $$\Rightarrow x = 1 + \frac{1}{x}.$$

   Next, we solve for x,

   $$0 = x - \left(1 + \frac{1}{x}\right) = \frac{x^2 - x - 1}{x}$$
   $$\Rightarrow 0 = x^2 - x - 1$$
   $$\therefore x = \frac{-(-1) \pm \sqrt{(-1)^2 - 4(-1)}}{2}$$
   $$= \frac{1 \pm \sqrt{5}}{2}.$$

   Since the "Golden Ratio" is defined as the ratio of lengths (positive quantities), it has to be positive. Hence we only accept the positive solution, $\phi = x = \frac{1+\sqrt{5}}{2}$.

   _____

2. (a) The displacement relative to the ground is given by $y(t) = y_0 + ut - \frac{gt^2}{2}$. With $u = 5$ m/s, $g = 10$ m/s^2 and $y_0 = 10$ m,

   $$y(t) = 10 + 5t - 5t^2.$$

   When the ball reaches the ground again, $y(t) = 0$, so

   $$0 = 10 + 5t - 5t^2$$
   $$\Rightarrow 0 = (t+1)(-5t + 10)$$
   $$\therefore t = -1 \text{ s or } 2 \text{ s}.$$

   Since t represents the time, which is positive, the only acceptable answer is $t = 2$ s.

   (b) Let us complete the square,

   $$\begin{aligned} y(t) &= -5t^2 + 5t + 10 \\ &= -5(t^2 - t - 2) \\ &= -5\left[(t - 1/2)^2 - 1/4 - 2\right] \\ &= -5\left[(t - 1/2)^2 - 9/4\right] \\ &= -5(t - 1/2)^2 + 45/4. \quad (2.1) \end{aligned}$$

   We see that the first term on the right hand side of (2.1) is never positive, $-5(t-1/2)^2 \leq 0$. So, $y(t)$ will reach the maximum height when this term vanishes. The maximum height is $y_{\max} = 45/4$ m and it happens at $t = 1/2$ s.

   (c) The ball starts at $t = 0$ from a platform which is 10 m above ground. It will move upwards but will eventually fall towards the ground. To find when it reaches $y = 5$ m, we solve

   $$\begin{aligned} 5 &= -5t^2 + 5t + 10 \\ 0 &= -5t^2 + 5t + 10 - 5 \\ 0 &= -5t^2 + 5t + 5 = -5(t^2 - t - 1) \\ \Rightarrow 0 &= t^2 - t - 1 \\ \therefore t &= \frac{-(-1) \pm \sqrt{1 - 4(-1)}}{2} \\ &= \frac{1 \pm \sqrt{5}}{2}. \end{aligned}$$

   Since $t \geq 0$, we have to choose $t = (1+\sqrt{5})/2$ s.

   (d) On the Moon, the gravitational strength is such that $g_{\text{moon}} = g_{\text{earth}}/6 = 5/3$ m/s^2. The new displacement equation is

   $$y(t) = 10 + 5t - \frac{5t^2}{6}. \quad (2.2)$$

   Let us redo parts (a, b) and (c) with this new acceleration.

   (i) When the ball reaches the ground, $y(t) = 0$.

So

$$
\begin{aligned}
0 &= 10 + 5t - \frac{5t^2}{6} \\
0 &= t^2 - 6t - 12 \\
t &= \frac{6 \pm \sqrt{36 - 4(-12)}}{2} \\
&= \frac{6 \pm \sqrt{84}}{2} = \left(3 \pm \sqrt{21}\right) \text{ s.}
\end{aligned}
$$

Since $t \geq 0$ only one answer is accepted which is $t = \left(3 + \sqrt{21}\right)$ s.

(ii) By completing the square,

$$
\begin{aligned}
y(t) &= -\frac{5t^2}{6} + 5t + 10 \\
&= -\frac{5}{6}\left(t^2 - 6t - 12\right) \\
&= -\frac{5}{6}\left[(t-3)^2 - 9 - 12\right] \\
&= -\frac{5}{6}\left[(t-3)^2 - 21\right] \\
&= -\frac{5}{6}(t-3)^2 + \frac{35}{2}.
\end{aligned}
$$

Therefore the ball reaches its maximum height $y_{\max} = 35/2$ m when $t = 3$ s.

(iii) Substitute $y = 5$ m into (2.2),

$$
\begin{aligned}
5 &= 10 + 5t - \frac{5t^2}{6} \\
0 &= 5 + 5t - \frac{5t^2}{6} \\
\Rightarrow 0 &= t^2 - 6t - 6 \\
\therefore t &= \frac{-(-6) \pm \sqrt{36 - 4(-6)}}{2} \\
&= \frac{-(-6) \pm \sqrt{60}}{2} = 3 \pm \sqrt{15}.
\end{aligned}
$$

Since $t \geq 0$, we have to choose $t = \left(3 + \sqrt{15}\right)$ s.

_____

3. (a)

$$
\begin{aligned}
2x + 2\sqrt{x} - 1 &= 0 \\
\Rightarrow 2\sqrt{x} &= 1 - 2x. \qquad (2.3)
\end{aligned}
$$

We can square both sides of the above equation to obtain,

$$
\begin{aligned}
4x &= (1 - 2x)^2 = 4x^2 - 4x + 1 \\
0 &= 4x^2 - 8x + 1 \\
x &= \frac{8 \pm \sqrt{64 - 4(4)}}{2(4)} = \frac{8 \pm \sqrt{48}}{8} \\
&= \frac{8 \pm 4\sqrt{3}}{8} = 1 \pm \sqrt{3}/2.
\end{aligned}
$$

We have to check the validity of the answer obtained since we have taken the square of both sides of the original equation (see page 9 of the book). Substituting $x = 1 + \sqrt{3}/2$ into equation (2.3) gives a negative value for the right-hand side, which is inconsistent with the positive left hand side. Therefore, only $x = 1 - \sqrt{3}/2$ is the valid solution.

(b)

$$
\begin{aligned}
0 &= 1 + \frac{3}{\sqrt{x}} - \frac{1}{x} \\
0 &= \frac{x + 3\sqrt{x} - 1}{x} \\
\Rightarrow 0 &= x + 3\sqrt{x} - 1 .
\end{aligned}
$$

Next, we move the $3\sqrt{x}$ term to the left hand side and then square both sides,

$$
\begin{aligned}
(-3\sqrt{x})^2 &= (x - 1)^2 \\
9x &= x^2 - 2x + 1 \\
\Rightarrow 0 &= x^2 - 11x + 1 \\
x &= \frac{11 \pm \sqrt{121 - 4}}{2} = \frac{11 \pm \sqrt{117}}{2} \\
&= \frac{11 \pm \sqrt{3^2 \times 13}}{2} = \frac{11 \pm 3\sqrt{13}}{2} .
\end{aligned}
$$

By substituting the answers into the original equation, we see that only $x = (11 - 3\sqrt{13})/2$ is valid.

(c)

$$
\begin{aligned}
3 &= \frac{1}{1-x} + \frac{1}{x+2} \\
0 &= \frac{1}{1-x} + \frac{1}{x+2} - 3 \\
0 &= \frac{(x+2) + (1-x) - 3(1-x)(x+2)}{(1-x)(x+2)} \\
0 &= \frac{(x+2) + (1-x) - 3(-x^2 - x + 2)}{(1-x)(x+2)} \\
0 &= \frac{3x^2 + 3x - 3}{(1-x)(x+2)} .
\end{aligned}
$$

The numerator must be zero, therefore

$$
\begin{aligned}
0 &= x^2 + x - 1 \\
x &= \frac{-1 \pm \sqrt{1 - 4(-1)}}{2} = (-1 \pm \sqrt{5})/2.
\end{aligned}
$$

_____

4. (a)

$$
\begin{aligned}
2 &\geq 5x + x^2 \\
0 &\geq x^2 + 5x - 2.
\end{aligned}
$$

Let $f_1(x) = x^2 + 5x - 2$. Its roots are

$$x = \frac{-5 \pm \sqrt{25 - 4(-2)}}{2} = \frac{-5 \pm \sqrt{33}}{2}.$$

Let $x_2 = (-5 + \sqrt{33})/2$ and $x_1 = (-5 - \sqrt{33})/2$, so we obtain (see Fig. 2.1),

$$0 \geq (x - x_1)(x - x_2)$$
$$\Rightarrow x_1 \leq x \leq x_2$$
$$\therefore \quad \frac{-5 - \sqrt{33}}{2} \leq x \leq \frac{\sqrt{33} - 5}{2}.$$

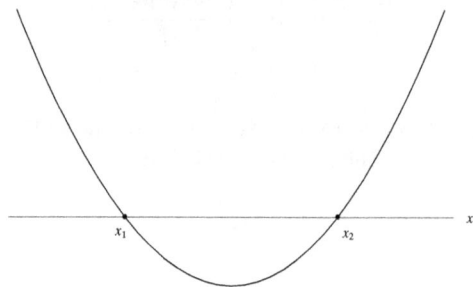

Figure 2.1: Intersection of parabola with x-axis.

(b)

$$4x \geq (x-1)(x-3)$$
$$0 \geq (x-1)(x-3) - 4x = x^2 - 8x + 3$$
$$\text{Let} \quad f_2(x) = x^2 - 8x + 3.$$

The roots of $f_2(x)$ are

$$x = \frac{8 \pm \sqrt{64 - 4(3)}}{2} = \frac{8 \pm \sqrt{52}}{2}$$
$$= 4 \pm \sqrt{13}.$$

Let $x_2 = 4 + \sqrt{13}$ and $x_1 = 4 - \sqrt{13}$ and thus we obtain (see Fig. 2.1),

$$0 \geq (x - x_1)(x - x_2)$$
$$\Rightarrow x_1 \leq x \leq x_2$$
$$\therefore \quad 4 - \sqrt{13} \leq x \leq 4 + \sqrt{13}.$$

(c)

$$2x + 5 \leq (3 - x)^2$$
$$0 \leq (9 - 6x + x^2) - (2x + 5)$$
$$\text{Let} \quad f_3(x) = x^2 - 8x + 4.$$

The roots of $f_3(x)$ are

$$x = \frac{8 \pm \sqrt{64 - 4(4)}}{2} = \frac{8 \pm \sqrt{48}}{2}$$
$$= \frac{8 \pm \sqrt{4^2 \times 3}}{2} = 4 \pm 2\sqrt{3}.$$

Let $x_2 = 2(2 + \sqrt{3})$ and $x_1 = 2(2 - \sqrt{3})$ and thus we obtain, (see figure 2.1)

$$0 \leq (x - x_1)(x - x_2)$$
$$\Rightarrow x \geq x_2 \text{ or } x \leq x_1$$
$$\therefore \quad x \geq 2(2 + \sqrt{3}) \text{ or } x \leq 2(2 - \sqrt{3}).$$

-----

5. *Note: Given $ax^2 + bx + c = 0$, the roots α and β satisfy $\alpha + \beta = -b/a$ and $\alpha\beta = c/a$.*

(a) Therefore,

$$\alpha + \beta = 7/2.$$
$$\alpha\beta = 1/2.$$
$$\frac{1}{\alpha} + \frac{1}{\beta} = \frac{\alpha + \beta}{\alpha\beta} = \frac{7/2}{1/2} = 7.$$

(b)

$$x = \frac{7 \pm \sqrt{49 - 4(2)}}{2(2)} = (7 \pm \sqrt{41})/4.$$

To verify the answer to part(a), we let $\alpha = (7 + \sqrt{41})/4$ and $4\ \beta = (7 - \sqrt{41})/4$. Then

$$\alpha + \beta = (7 + \sqrt{41})/4 + (7 - \sqrt{41})/4 = 7/2 ;$$
$$\alpha\beta = \frac{7 + \sqrt{41}}{4} \times \frac{7 - \sqrt{41}}{4} ;$$
$$= \frac{49 - 41}{16} = \frac{8}{16} = \frac{1}{2} ;$$

We see that the result is the same as in part(a).

-----

6. Given $f(x) = x^2 + bx + c$.
(a) If $f(x)$ has a double root at $x = 5$, it implies that $f(x) = a(x - 5)^2 = a(x^2 - 10x + 25)$, with a a constant. Comparing with the given expression, we deduce that $b = -10$, $c = 25$.

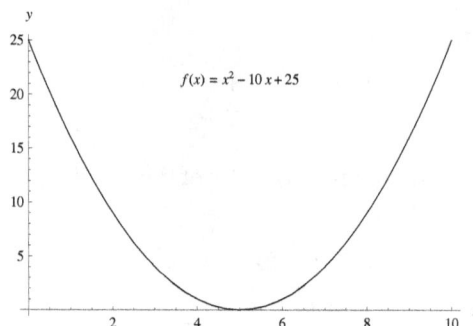

Figure 2.2: Plot of $f(x) = x^2 - 10x + 25$. There is a double root at $x = 5$.

(b) If $f(x)$ has roots at $x = -1$ and $x = 2$, it must be of the form $f(x) = a(x+1)(x-2) = a(x^2 - x - 2)$, with a a constant. Therefore, $b = -1$, $c = -2$.

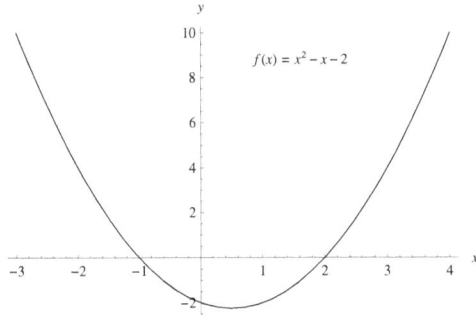

Figure 2.3: Plot of $f(x) = x^2 - x - 2$. The roots are at $x = -1$; $x = 2$.

7. (a) Since $f(x) = x^2 - 10x + 25 = (x-5)^2$ is always positive definite, the modulus graph is exactly the same as the original graph, $|f(x)| = f(x)$ (see Fig.2.2). From the plot, the range that y takes for x between 0 and the positive root $x = 5$ is $0 \le y \le 25$.

(b)

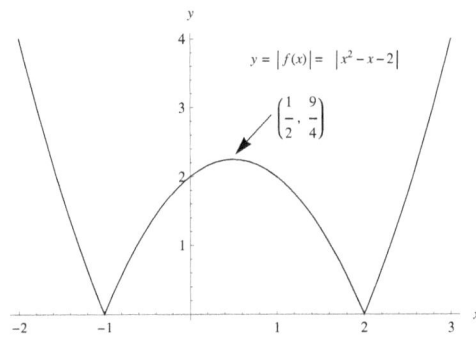

Figure 2.4: Plot of $|f(x)|$ where $f(x) = x^2 - x - 2$.

The local maximum of the graph $|f(x)|$ can be determined as follows,

$$f(x) = |x^2 - x - 2| = |(x-1/2)^2 - 1/4 - 2|$$
$$= |(x-1/2)^2 - 9/4|.$$

Thus the maximum occurs at $x = 1/2$ and $y_{\max} = 9/4$. The range that y takes for x between 0 and the positive root $x = 2$ is $0 \le y \le 9/4$.

8. *Recall that a quadratic equation $Ax^2 + Bx + C = 0$ has discriminant $\Delta = B^2 - 4AC$. The discriminant of $f(x) = x^2 + bx + 2b$ is $b^2 - 8b = b(b-8)$.*

(a) For $f(x)$ to have no real roots, the discriminant must be negative,

$$0 > b(b-8)$$
$$\therefore \quad 0 < b < 8.$$

(b) If $f(x)$ has a root at $x = -1$, then $f(-1) = 0$, implying

$$0 = 1^2 + b(-1) + 2b$$
$$\Rightarrow b + 1 = 0$$
$$\therefore b = -1.$$

(c) For distinct real roots, the discriminant must be positive

$$0 < b(b-8)$$
$$\Rightarrow \quad b < 0 \quad \text{or} \quad b > 8.$$

Furthermore, if one of the roots, α, is positive while the other root β is negative, then $\alpha\beta < 0$. This condition imposes $\alpha\beta = C/A = 2b < 0 \Rightarrow b < 0$. To fulfil both constraints, we need to choose $b < 0$.

(d) As in part (c), we need to impose

$$0 < b(b-8)$$
$$\Rightarrow \quad b < 0 \quad \text{or} \quad b > 8$$

to obtain two distinct real roots. Furthermore, if both roots are negative then $\alpha\beta = 2b > 0 \Rightarrow b > 0$. To fulfil both conditions, we need to choose $b > 8$.

9. (a) Given that $\alpha + \beta = 2$ and $\alpha\beta = -2$, we can substitute $\beta = -2/\alpha$ into the first equation,

$$2 = \alpha + \left(\frac{-2}{\alpha}\right)$$
$$2\alpha = \alpha^2 - 2$$
$$\Rightarrow 0 = \alpha^2 - 2\alpha - 2$$
$$\therefore \alpha = \frac{2 \pm \sqrt{4 - 4(-2)}}{2}$$
$$= \frac{2 \pm \sqrt{12}}{2} = 1 \pm \sqrt{3}.$$

Therefore the two roots are $1 \pm \sqrt{3}$. (You can verify by substitution that if $\alpha = 1 + \sqrt{3}$, then $\beta = 1 - \sqrt{3}$, and vice versa.)

(b) Once we know the roots, we can factorise the polynomial,

$$f(x) = A(x - \alpha)(x - \beta)$$

where A is a constant. By comparing coefficients of the x^2 term on both sides, we deduce that $A = 3$. Therefore

$$
\begin{aligned}
f(x) &= 3(x - \alpha)(x - \beta) \\
&= 3x^2 - 3(\alpha + \beta)x + 3\alpha\beta \\
&= 3x^2 - 3(2)x + 3(-2) \\
&= 3(x^2 - 2x - 2).
\end{aligned}
$$

(c) As in part (b),

$$f(x) = a(x^2 - 2x - 2).$$

For the intersection point, we have to solve the simultaneous equations,

$$
\begin{aligned}
y &= x - 3a = a(x^2 - 2x - 2) \\
\Rightarrow 0 &= ax^2 - (2a + 1)x + a.
\end{aligned}
$$

Given that $f(x)$ meets the line $y = x - 3a$ at only one point, the quadratic equation above must have a unique solution. This implies that the discriminant of that quadratic equation must vanish,

$$
\begin{aligned}
0 &= (-2a - 1)^2 - 4a^2 \\
&= 4a + 1 \\
\therefore a &= -1/4.
\end{aligned}
$$

---

10. Given that the cubic equation $f(x) = x^3 - ax^2 - x + 3$ has a root at $x = -1$, therefore $f(-1) = 0 = -1 - a + 1 + 3$. So $a = 3$. By long division,

$$
\begin{array}{r}
x^2 - 4x + 3 \\
x + 1 \overline{)\ x^3 - 3x^2 - x + 3} \\
\underline{-(x^3 + x^2)} \\
-4x^2 - x + 3 \\
\underline{-(-4x^2 - 4x)} \\
3x + 3 \\
\underline{3x + 3} \\
0\ .
\end{array}
$$

So, $f(x) = (x + 1)(x^2 - 4x + 3)$ and the other roots of $f(x)$ are the roots of the quadratic

equation. After factorising the quadratic part, we have $f(x) = (x + 1)(x - 1)(x - 3)$.

(b) Sketch of the curve

Figure 2.5: Plot of $f(x) = (x + 1)(x - 1)(x - 3)$.

(c) Substitute $y = x^3 - c$ into $y = f(x)$ to find the intersection points. We have

$$
\begin{aligned}
x^3 - c &= x^3 - 3x^2 - x + 3 \\
0 &= 3x^2 + x - (c + 3).
\end{aligned}
$$

Imposing the discriminant condition $B^2 - 4AC \geq 0$ to obtain real solutions of x,

$$
\begin{aligned}
1^2 - 4(3)(-c - 3) &\geq 0 \\
\Rightarrow 12c + 37 &\geq 0 \\
\therefore c &\geq -37/12.
\end{aligned}
$$

---

11. (a)

$$
\begin{aligned}
\frac{2x + 1}{x^2 + 3x + 2} &= \frac{2x + 1}{(x + 1)(x + 2)} \\
&\equiv \frac{A}{(x + 1)} + \frac{B}{(x + 2)}.
\end{aligned}
$$

Cross-multiplication gives

$$2x + 1 = A(x + 2) + B(x + 1).$$

One can solve for the constant A by substituting $x = -1$, which gives $A = -1$. Similarly, by setting $x = -2$ we obtain $B = 3$. The partial fraction form is therefore

$$\frac{2x + 1}{x^2 + 3x + 2} = \frac{3}{(x + 2)} - \frac{1}{(x + 1)}.$$

(b) First we use long division to reduce the degree of the numerator below that of the denominator,

$$
\begin{array}{r}
5 \\
x^2 + 3x + 2 \overline{)\ 5x^2 + 2x + 1} \\
\underline{-(5x^2 + 15x + 10)} \\
-13x - 9
\end{array}
$$

so

$$\frac{5x^2 + 2x + 1}{x^2 + 3x + 2} = 5 + \frac{-13x - 9}{(x+1)(x+2)}.$$

The fractional part may be split into partial fractions. We expect

$$\frac{-13x - 9}{(x+1)(x+2)} \equiv \frac{A}{(x+1)} + \frac{B}{(x+2)}.$$

Cross-multiplication gives

$$-13x - 9 \equiv A(x+2) + B(x+1).$$

By setting $x = -1$ we obtain $A = -13(-1) - 9 = 4$. Also one can obtain $B = -17$ by setting $x = -2$. As a result,

$$\frac{5x^2 + 2x + 1}{x^2 + 3x + 2} = 5 + \frac{4}{(x+1)} - \frac{17}{(x+2)}.$$

(c) Proceeding as in part (b), a long division gives

$$
\begin{array}{r}
5x - 3 \\
x + 1 \overline{)\, 5x^2 + 2x + 1} \\
-(5x^2 + 5x) \\
\hline
-3x + 1 \\
-(-3x - 3) \\
\hline
4\,.
\end{array}
$$

so we have

$$\frac{5x^2 + 2x + 1}{x + 1} = 5x - 3 + \frac{4}{x + 1}.$$

2.2 Solutions to Odd-Numbered Problems

1. (a) Given that $xy = 1$ m^2 and also

$$\frac{x}{y} = \frac{y}{x/2}$$

$$\Rightarrow x^2 = 2y^2 = \frac{2}{x^2}$$

$$\therefore x = 2^{1/4}; \quad y = 2^{-1/4}\,.$$

(b) The dimension of the papers are

$$
\begin{aligned}
A_1 &= x/2 \times y \\
A_2 &= x/2 \times y/2 \\
A_3 &= x/4 \times y/2 \\
A_4 &= x/4 \times y/4.
\end{aligned}
$$

The dimension of A_4 paper is $x/4 = 2^{1/4}/4 = 0.297$m $= 297$mm. Also, $y/4 = 2^{-1/4}/4 = 0.210$m $= 210$mm.

3. Given $y = ax - bx^2$, if the ball lands at $x = 25$, $y = -25/2$, then $a = 25b - 1/2$. Next, when $y = 0$ we have either $x = 0$ (initial position) or $x = a/b$, thus at its maximum height, the x-coordinate of the ball is $x = a/(2b)$ (the curve is a parabola which is symmetrical).

Substitute all this information into the trajectory equation to obtain,

$$
\begin{aligned}
y_{max} &= 100 = a\left(\frac{a}{2b}\right) - b\left(\frac{a}{2b}\right)^2 = \frac{a^2}{4b} \\
\Rightarrow 400b &= \left(25b - \frac{1}{2}\right)^2 \\
0 &= 625b^2 - 425b + \frac{1}{4} \\
b &= \frac{17 \pm 12\sqrt{2}}{50}.
\end{aligned}
$$

By direct substitution, we get

$$
\begin{aligned}
a &= 25b - \frac{1}{2} \\
&= 8 \pm 6\sqrt{2}.
\end{aligned}
$$

However, since both a and b are positive, we need to choose the "$+$" solution for a and b above. We leave the sketch to you.

(b) The x-coordinate of the maximum point in the trajectory is

$$
\begin{aligned}
x &= \frac{a}{2b} = \frac{8 + 6\sqrt{2}}{2(17 + 12\sqrt{2})/50} \\
&= \frac{50(4 + 3\sqrt{2})}{(17 + 12\sqrt{2})} \times \frac{17 - 12\sqrt{2}}{17 - 12\sqrt{2}} \\
&= -200 + 150\sqrt{2}.
\end{aligned}
$$

So, the maximum point has coordinates $(-200 + 150\sqrt{2},\ 100)$.

(c) At $y = 0$ we have $x = 0$ or

$$x = \frac{a}{b} = 100(3\sqrt{2} - 4).$$

(d) For $y \geq 50$, we have

$$
\begin{aligned}
50 &\leq ax - bx^2 \\
0 &\geq x^2 - \frac{a}{b}x + \frac{50}{b} \\
\Rightarrow 0 &\geq (x - x_+)(x - x_-)
\end{aligned}
$$

where

$$
\begin{aligned}
x_{\pm} &= \frac{a/b \pm \sqrt{a^2/b^2 - 4(50/b)}}{2} \\
&= 20.71 \text{ or } 3.55
\end{aligned}
$$

So, the allowed domain is $3.55 \leq x \leq 20.71$.

5. (a)
$$y = 2x^2 - 3x + 3 = x + 2$$
$$0 = 2x^2 - 4x + 1$$
$$x = \frac{4 \pm \sqrt{16 - 4(2)}}{4} = 1 \pm \frac{1}{\sqrt{2}}.$$

Correspondingly, $y = x + 2 = 3 \pm (1/\sqrt{2})$.

(b)
$$y = x^2 - x - 1 = 3x^2 - 4$$
$$\Rightarrow x = \frac{-1 \pm \sqrt{1 - 4(2)(-3)}}{4}$$
$$= 1 \text{ or } -3/2.$$

Correspondingly, $y = -1$ or $11/4$.

(c)
$$y^2 = 2x^2 - 3x + 3 = (-x + 2)^2$$
$$\Rightarrow x = \frac{-1 \pm \sqrt{1 - 4(-1)}}{2} = \frac{-1 \pm \sqrt{5}}{2}.$$

Correspondingly $y = -x + 2 = (5 \mp \sqrt{5})/2$.

7. Given the quadratic equation $2x^2 - 7x + p = 0$ with roots α, β such that $\alpha > \beta$. We have $\alpha + \beta = 7/2$ and $\alpha\beta = p/2$.

(a)
$$\alpha - \beta = \sqrt{(\alpha + \beta)^2 - 4\alpha\beta}$$
$$= \sqrt{(7/2)^2 - 4(p/2)} = \frac{\sqrt{49 - 8p}}{2}.$$

We chose only the positive square-root solution since $\alpha > \beta$.

(b)
$$\frac{\beta}{\alpha} + \frac{\alpha}{\beta} = \frac{\alpha^2 + \beta^2}{\alpha\beta} = \frac{(\alpha + \beta)^2 - 2\alpha\beta}{\alpha\beta}$$
$$= \frac{(7/2)^2 - 2(p/2)}{p/2} = \frac{49}{2p} - 2.$$

(c)
$$\alpha^4 + \beta^4 = (\alpha^2 + \beta^2)^2 - 2\alpha^2\beta^2$$
$$= \left(\frac{49}{4} - p\right)^2 - \frac{p^2}{2}$$
$$= \frac{8p^2 - 392p + 2401}{16}.$$

9. The sides of the rectangle are a and $1/a$. The perimeter is $P = 2a + 2/a$. The diagonal is $D = \sqrt{a^2 + 1/a^2}$. Since $D = 3P/8$, therefore
$$\sqrt{a^2 + 1/a^2} = \frac{3}{4}(a + 1/a)$$
$$16a^2 + 16/a^2 = 9(a^2 + 1/a^2 + 2)$$
$$\Rightarrow 0 = 7a^2 - 18 + 7/a^2$$

Multiply both sides by a^2 and let $x = a^2$,
$$0 = 7x^2 - 18x + 7$$
$$x = \frac{18 \pm \sqrt{18^2 - 4(7)(7)}}{2(7)} = \frac{9 \pm 4\sqrt{2}}{7}$$
$$a = \sqrt{\frac{9 \pm 4\sqrt{2}}{7}} = 1.45 \text{ or } 0.69.$$

11. (a) For $h = 2R$ (fully filled), $V(h = 2R) = \frac{\pi(2R)^2}{3}[3R - (2R)] = \frac{4\pi}{3}R^3$. This is precisely the volume of a solid sphere with radius R.

(b) If $R = 10$ and $V = \frac{29\pi}{3}$,
$$\frac{\pi h^2}{3}(30 - h) = \frac{29\pi}{3}$$
$$\Rightarrow f(h) = h^3 - 30h^2 + 29 = 0.$$

It is obvious that $h = 1$ is the solution of above equation.

(c) Since $h_1 = 1$ is one of the roots, we can write $f(h) = (h - 1)g(h)$. By long division, we get $g(h) = h^2 - 29h - 2$. The other roots of $f(h)$ can be obtained by requiring $g(h) = 0 \Rightarrow h = \frac{29 \pm \sqrt{957}}{2}$. Numerically, they are $h_2 = 29.968, h_3 = -0.968$.

The height is constrained by the diameter of the sphere, $0 < h \leq 2R$. So, only $h_1 = 1$ is a physical solution.

13. (a) Given that $P(x) = x^4 + ax^3 + bx^2 + 3$ is identical to $S(x) = x(x+1)(x-2)Q(x)+c+1$, we can evaluate both expressions for particular values of x and compare the results. When $x = 0$ we have $P(0) = 3$ and $S(0) = c + 1$, thus $c = 2$. Next by letting $x = -1$, $P(-1) = S(-1) \Rightarrow a - b = 1$. Also, $P(2) = S(2) \Rightarrow 2a + b = -4$. Solving the two simultaneous equations gives $a = -1, b = -2$.

(b) With $P(x) = x^4 - x^3 - 2x^2 + 3$ and $S(x) = x(x+1)(x-2)Q(x) + 3$, direct comparison (or long division) gives $Q(x) = x$.

(c) By the remainder theorem, the remainder is

$$P(1) = (1)^4 - (1)^3 - 2(1)^2 + 3 = 1.$$

15. The general form of the polynomial $g(x)$ (with roots $x = 1, p, 2p$) is

$$g(x) = a(x-1)(x-p)(x-2p), \quad a \neq 0.$$

Also, since the coefficient of x^0 in $g(x)$ is 5, this implies $-2ap^2 = 5$. As the remainder is -2 when divided by $(x-2)$, therefore $g(2) = a(2-p)(2-2p) = -2$. With these pieces of information, we can find the possible values for p,

$$-2 = -\frac{5}{2p^2}(2-p)(2-2p)$$
$$0 = 3p^2 - 15p + 10$$
$$\Rightarrow p = \frac{15 \pm \sqrt{105}}{6}.$$

---

17. (a) Perform a long division and rewrite the expression as

$$\frac{x^4 + 1}{x^3 + x} = x + \frac{1 - x^2}{x(x^2 + 1)}$$
$$\equiv x + \frac{A}{x} + \frac{Bx + C}{x^2 + 1}.$$

We have

$$A(x^2 + 1) + (Bx + C)x = 1 - x^2.$$

Substituting $x = 0$ gives $A = 1$. Also $x = 1$ gives $B + C = -2$; $x = -1$ gives $-B + C = 2$. Solving both equations, we obtain $C = 0$ and $B = -2$. Thus,

$$\frac{x^4 + 1}{x^3 + x} = x + \frac{1}{x} - \frac{2x}{x^2 + 1}.$$

(b) After a long division, we have

$$\frac{5x^2 + 2x + 1}{x^2 + 2x + 1} = 5 - \frac{8x + 4}{x^2 + 2x + 1}$$
$$\equiv 5 + \frac{A}{x+1} + \frac{B}{(x+1)^2}.$$

We have to solve

$$A(x+1) + B = -(8x + 4).$$

Substituting $x = -1$ gives $B = 4$ and setting $x = 0$ gives $A = -8$. Thus we have

$$\frac{5x^2 + 2x + 1}{x^2 + 2x + 1} = 5 - \frac{8}{x+1} + \frac{4}{(x+1)^2}.$$

(c) Rewrite as

$$\frac{5x^2 + 2x + 1}{(x^2 + 2x + 1)(x + 2)}$$
$$= \frac{A}{x+2} + \frac{B}{x+1} + \frac{C}{(x+1)^2}.$$

We need to determine the constants in

$$A(x+1)^2 + B(x+1)(x+2) + C(x+2)$$
$$= 5x^2 + 2x + 1.$$

Substituting $x = -2$ gives us $A = 17$ while $x = -1$ gives us $C = 4$. Finally setting $x = 0$ gives $A + 2B + 2C = 1$ which translates into $B = -12$. Thus,

$$\frac{5x^2 + 2x + 1}{(x^2 + 2x + 1)(x + 2)}$$
$$= \frac{17}{x+2} - \frac{12}{x+1} + \frac{4}{(x+1)^2}.$$

(d) Rewrite as

$$\frac{5x^2 + 2x + 1}{(x^2 + 1)(x + 2)} = \frac{Ax + B}{x^2 + 1} + \frac{C}{x+2}.$$

We need to solve

$$(Ax + B)(x + 2) + C(x^2 + 1) = 5x^2 + 2x + 1.$$

Substituting $x = -2$ gives us $C = 17/5$ while $x = 0$ gives us $2B + C = 1 \Rightarrow B = -6/5$. Finally let $x = -1$ which gives $-A + B + 2C = 4$ and hence $A = 8/5$. Thus,

$$\frac{5x^2 + 2x + 1}{(x^2 + 1)(x + 2)} = \frac{8x - 6}{5(x^2 + 1)} + \frac{17}{5(x + 2)}.$$

---

19. Write $1/(k^2 - 1)$ in the partial fraction form $\frac{1}{2}\left(\frac{1}{k-1} - \frac{1}{k+1}\right)$. Then, on writing out the terms explicitly, and re-arranging, the given series is seen to be

$$\frac{1}{2}\left(1 + \frac{1}{2}\right) + R - R$$

where $R = \frac{1}{2}\left(\frac{1}{3} + \frac{1}{4} + \frac{1}{5} + ...\right)$.

Most terms cancel, leaving the result $3/4$.

Chapter 3

Simultaneous Equations and Matrices

3.1 Solutions to Exercises

1. Denote the number of passengers in boats A and B by a and b respectively. From the question, we have the following information

$$a + 1 = 2(b - 1) ;$$
$$a - 1 = b + 1.$$

Taking the difference between the two equations, we get $(a+1)-(a-1)=2(b-1)-(b+1) \Rightarrow 2 = b - 3$. Thus we obtain $b = 5$. After substitution, $a = 2(b-1)-1 = 2(5-1)-1 = 7$.

_____

2. Denote the number of teams with three students and the number with five students by x and y respectively. Since we have in total 200 students, $3x + 5y = 200$, or $y = (200 - 3x)/5$. Also, we are told that total number of teams formed is less than 50, that is $x + y < 50$. By substituting y into the inequality we get

$$50 > x + \left(\frac{200 - 3x}{5}\right)$$
$$50 > \frac{5x + 200 - 3x}{5}$$
$$250 > 5x + (200 - 3x)$$
$$250 - 200 > 5x - 3x$$
$$50 > 2x$$
$$\therefore 25 > x.$$

So, the number of teams with three members must be less than 25.

_____

3. (a) Denote the two numbers by x and y. Given that their sum is 10, $x+y = 10$. Also, $\frac{1}{x} + \frac{1}{y} = 2$, which we can re-write it as $x+y = 2xy$. We can substitute the first condition $y = 10 - x$

into the second to solve for y. Explicitly,

$$x + (10 - x) = 2x(10 - x)$$
$$10 = 20x - 2x^2$$
$$-10 = 2x^2 - 20x$$
$$0 = 2x^2 - 20x + 10$$
$$0 = x^2 - 10x + 5$$
$$\therefore x = \frac{10 \pm \sqrt{100 - 4(1)(5)}}{2}$$
$$= \frac{10 \pm \sqrt{80}}{2}$$
$$= \frac{10 \pm \sqrt{4^2 \times 5}}{2} = 5 \pm 2\sqrt{5}.$$

When $x_1 = 5 + 2\sqrt{5}$, we have $y_1 = 10 - (5 + 2\sqrt{5}) = 5 - 2\sqrt{5}$. The other solution is $x_2 = 5 - 2\sqrt{5}$ and $y_2 = 5 + 2\sqrt{5}$.

_____

4. We denote the two numbers by x and y. We have

$$x + y = 18 , \qquad (3.1)$$
$$\frac{1}{x} + \frac{1}{y} > 4. \qquad (3.2)$$

Eq.(3.2) can be rewritten as $\frac{x+y}{xy} > 4$. Replacing y in this equation by $y = 18 - x$ (from the first equation), and simplifying, we get

$$\frac{18}{x(18 - x)} > 4$$
$$\Rightarrow 2x^2 - 36x - 9 > 0 . \qquad (3.3)$$

As the roots of $2x^2 - 36x - 9 = 0$ are $x = (18 \pm 3\sqrt{34})/2$, then $y = 18 - x = (18 \mp 3\sqrt{34})/2$.

The range of x satisfying the inequality (3.3) can be found as in Exercise (4c) of the previous chapter. The constraint on the smaller number is therefore $0 < x < \frac{18 - 3\sqrt{34}}{2}$.

_____

5. Mathematically, this is equivalent to solving the set of simultaneous equations involving $y = ax - bx^2$ and $y = mx$, the latter being the inclined line representing the platform of slope m.

---

6. *Note: For matrix* $\begin{pmatrix} a & b \\ c & d \end{pmatrix}$, *its determinant is* $\det(M) = ad - bc$. *The inverse* $M^{-1} = \frac{1}{\det(M)} \begin{pmatrix} d & -b \\ -c & a \end{pmatrix}$.

(a) Given the matrix $A = \begin{pmatrix} 1 & 2 \\ 3 & 4 \end{pmatrix}$, $\det A = ad - bc = (1)(4) - (2)(3) = -2$. The inverse matrix is

$$A^{-1} = \frac{1}{-2}\begin{pmatrix} 4 & -2 \\ -3 & 1 \end{pmatrix} = \begin{pmatrix} -2 & 1 \\ 3/2 & -1/2 \end{pmatrix}.$$

Let us verify the result

$$AA^{-1} = \begin{pmatrix} 1 & 2 \\ 3 & 4 \end{pmatrix}\begin{pmatrix} -2 & 1 \\ 3/2 & -1/2 \end{pmatrix} = \begin{pmatrix} 1 & 0 \\ 0 & 1 \end{pmatrix};$$

$$A^{-1}A = \begin{pmatrix} -2 & 1 \\ 3/2 & -1/2 \end{pmatrix}\begin{pmatrix} 1 & 2 \\ 3 & 4 \end{pmatrix} = \begin{pmatrix} 1 & 0 \\ 0 & 1 \end{pmatrix}.$$

We see that A^{-1} is indeed the inverse matrix of A since their product produces the identity matrix.

(b) Given matrix $A = \begin{pmatrix} 1 & 2 \\ 2 & 4 \end{pmatrix}$, $\det A = ad - bc = (1)(4) - (2)(2) = 0$. Since this matrix is singular it has no inverse matrix.

(c) Given the matrix $A = \begin{pmatrix} -1 & 1 \\ -2 & -2 \end{pmatrix}$, $\det A = ad - bc = (-1)(-2) - (1)(-2) = 4$. The inverse matrix is

$$A^{-1} = \frac{1}{4}\begin{pmatrix} -2 & -1 \\ 2 & -1 \end{pmatrix} = \begin{pmatrix} -1/2 & -1/4 \\ 1/2 & -1/4 \end{pmatrix}.$$

Furthermore,

$$AA^{-1} = \begin{pmatrix} -1 & 1 \\ -2 & -2 \end{pmatrix}\begin{pmatrix} \frac{-1}{2} & \frac{-1}{4} \\ \frac{1}{2} & \frac{-1}{4} \end{pmatrix} = \begin{pmatrix} 1 & 0 \\ 0 & 1 \end{pmatrix};$$

$$A^{-1}A = \begin{pmatrix} \frac{-1}{2} & \frac{-1}{4} \\ \frac{1}{2} & \frac{-1}{4} \end{pmatrix}\begin{pmatrix} -1 & 1 \\ -2 & -2 \end{pmatrix} = \begin{pmatrix} 1 & 0 \\ 0 & 1 \end{pmatrix}.$$

(d) Given the matrix $A = \begin{pmatrix} -4 & 3 \\ -2 & 1 \end{pmatrix}$, $\det A = ad - bc = (-4)(1) - (3)(-2) = 2$. The inverse matrix is

$$A^{-1} = \frac{1}{2}\begin{pmatrix} 1 & -3 \\ 2 & -4 \end{pmatrix} = \begin{pmatrix} 1/2 & -3/2 \\ 1 & -2 \end{pmatrix}.$$

As expected,

$$AA^{-1} = \begin{pmatrix} -4 & 3 \\ -2 & 1 \end{pmatrix}\begin{pmatrix} \frac{1}{2} & \frac{-3}{2} \\ 1 & -2 \end{pmatrix} = \begin{pmatrix} 1 & 0 \\ 0 & 1 \end{pmatrix};$$

$$A^{-1}A = \begin{pmatrix} \frac{1}{2} & \frac{-3}{2} \\ 1 & -2 \end{pmatrix}\begin{pmatrix} -4 & 3 \\ -2 & 1 \end{pmatrix} = \begin{pmatrix} 1 & 0 \\ 0 & 1 \end{pmatrix}.$$

---

7. (a) We can re-arrange the inequality $-3x+3 > 2-x$ to obtain

$$\begin{aligned} -3x + 3 &> 2 - x \\ 3 - 2 &> -x + 3x \\ 1 &> 2x \\ \therefore \frac{1}{2} &> x. \end{aligned}$$

(b) We are given $x - 1 > 3x - 4 > -2 - x$. We can treat this as two simultaneous inequalities:

$$\begin{aligned} x - 1 &> 3x - 4 \; ; \\ 3x - 4 &> -2 - x. \end{aligned}$$

We can rearrange the terms and obtain two simplified inequalities,

$$\begin{aligned} 2x < 3 &\Rightarrow x < 3/2 \; ; \\ 4x > 2 &\Rightarrow x > 1/2. \end{aligned}$$

The common domain of these inequalities is $\frac{1}{2} < x < \frac{3}{2}$.

(c) We are given $5 - 2x > 2 - x > 3x - 4$. As in part (b), we have two simultaneous inequalities,

$$\begin{aligned} x &< 3 \; ; \\ 4x &< 6 \Rightarrow x < 3/2. \end{aligned}$$

The common domain is $x < \frac{3}{2}$.

8. (a) Given that $2y - 5x = 1$ and $2x - 5y = 3$, we can re-write the linear simultaneous equations in matrix form,

$$\begin{pmatrix} -5 & 2 \\ 2 & -5 \end{pmatrix}\begin{pmatrix} x \\ y \end{pmatrix} = \begin{pmatrix} 1 \\ 3 \end{pmatrix}$$

$$\Rightarrow \begin{pmatrix} x \\ y \end{pmatrix} = \begin{pmatrix} -5 & 2 \\ 2 & -5 \end{pmatrix}^{-1}\begin{pmatrix} 1 \\ 3 \end{pmatrix}.$$

We need to calculate the inverse matrix in order to determine the unknowns x and y. The determinant is $(-5)(-5) - 2(2) = 21$. The inverse matrix is

$$\begin{pmatrix} -5 & 2 \\ 2 & -5 \end{pmatrix}^{-1} = \frac{1}{21}\begin{pmatrix} -5 & -2 \\ -2 & -5 \end{pmatrix}$$

$$= \frac{-1}{21}\begin{pmatrix} 5 & 2 \\ 2 & 5 \end{pmatrix}.$$

Therefore

$$\begin{pmatrix} x \\ y \end{pmatrix} = \frac{-1}{21} \begin{pmatrix} 5 & 2 \\ 2 & 5 \end{pmatrix} \begin{pmatrix} 1 \\ 3 \end{pmatrix}$$

$$= \frac{-1}{21} \begin{pmatrix} 5(1) + 2(3) \\ 2(1) + 5(3) \end{pmatrix}$$

$$= \begin{pmatrix} -11/21 \\ -17/21 \end{pmatrix}.$$

(b) Given that $2y - 5x = 1$ and $y = 3x^2 - 2$, one can substitute the second equation into the first one, and solve the resulting quadratic equation,

$$\begin{aligned} 1 &= 2(3x^2 - 2) - 5x \\ &= 6x^2 - 5x - 4 \\ \Rightarrow 0 &= 6x^2 - 5x - 5 \\ \therefore x &= \frac{5 \pm \sqrt{25 - 4(6)(-5)}}{12} \\ &= \frac{5 \pm \sqrt{145}}{12}. \end{aligned}$$

Then we get y from $y = (5x + 1)/2$,

$$\begin{aligned} y &= \frac{\left(5 \times \frac{5 \pm \sqrt{145}}{12} + 1\right)}{2} \\ &= \frac{5(5 \pm \sqrt{145}) + 12}{24} \\ &= \frac{25 \pm 5\sqrt{145} + 12}{24} \\ &= \frac{37 \pm 5\sqrt{145}}{24}. \end{aligned}$$

(c) Given that $2y - 5\sqrt{x} = 1$ and $2x - 5y = 3$. From the first equation we have

$$5\sqrt{x} = 2y - 1$$

From this we see that $x \geq 0$ and $y \geq 1/2$. Write

$$x = \left(\frac{2y - 1}{5}\right)^2,$$

substitute it into the second equation and solve the resulting quadratic equation in y. Explicitly,

$$\begin{aligned} 3 &= 2\left(\frac{2y - 1}{5}\right)^2 - 5y \\ 0 &= \frac{2(2y - 1)^2 + (25)(-5y - 3)}{25} \\ \Rightarrow 0 &= 2(4y^2 - 4y + 1) - 125y - 75 \\ 0 &= 8y^2 - 133y - 73 \\ y &= \frac{133 \pm \sqrt{133^2 - 4(8)(-73)}}{2(8)} \\ &= \frac{133 \pm \sqrt{20025}}{16} = \frac{133 \pm 15\sqrt{89}}{16}. \end{aligned}$$

Since $y \geq 1/2$, we only accept $y = (133 + 15\sqrt{89})/16$. The corresponding x is

$$\begin{aligned} x &= \frac{3 + 5 \times \frac{133 + 15\sqrt{89}}{16}}{2} \\ &= \frac{48 + 5(133 + 15\sqrt{89})}{32} = \frac{713 + 75\sqrt{89}}{32}. \end{aligned}$$

(d) Given that $1/(y - \sqrt{x}) = 1/(\sqrt{x} + 2y) = -1$. We treat it as two equalities. Firstly, we re-write

$$\begin{aligned} \frac{1}{y - \sqrt{x}} &= \frac{1}{\sqrt{x} + 2y} \\ \Rightarrow y - \sqrt{x} &= \sqrt{x} + 2y \\ -2\sqrt{x} &= y \Rightarrow \sqrt{x} = -y/2. \end{aligned}$$

whereby we assume $y \neq \sqrt{x}$ and $y \neq -\sqrt{x}/2$ when taking the reciprocal. Next, we substitute $\sqrt{x} = -y/2$ into the second equality,

$$\begin{aligned} \frac{1}{\sqrt{x} + 2y} &= -1 \\ \sqrt{x} + 2y &= -1 \\ \Rightarrow -\frac{y}{2} + 2y &= -1 \\ \frac{3y}{2} &= -1 \\ \therefore y &= -2/3. \end{aligned}$$

Correspondingly, $x = y^2/4 = (\frac{-2}{3})^2/4 = 1/9$.

3.2   Solutions   to   Odd-Numbered Problems

1. Denote the number of teams with three (five) members as N_1 and N_2 respectively. The total number of teams is $N = N_1 + N_2$ and there are in total 200 students, that is, $3N_1 + 5N_2 = 200$. Since $N_1 > N_2$, we have

$$N_1 > \frac{200 - 3N_1}{5} \quad \Rightarrow \quad N_1 > 25$$
$$N_2 < \frac{200 - 5N_2}{3} \quad \Rightarrow \quad N_2 < 25.$$

Trying $N_1 = 26, 27, 28, 29$ yields no integer solutions for N_2. When $N_1 = 30$ we have the first integer solution for $N_2 = 22$ and thus $N = 52$. Since $N_2 = (200 - 3N_1)/5$, we have $N = \frac{2}{5}N_1 + 40$. Therefore $N = 52$ is the smallest value for the possible total number of teams formed.

_____

3. Denote the number of buses rented from company X and Y by N_x and N_y respectively. Also, let the rental price for the buses be p_x and p_y. For the same amount spent, we have the constraint $N_x p_x = N_y p_y$. Furthermore, given that $p_x = 5p_y$, we get $N_y = 5N_x$.

The other constraint is from the total number of passengers, $40N_x + 30N_y \geq 170$. We use " $\geq$ " instead of " $=$ " since the buses may not be fully occupied. By substitution, $40N_x + 30(5N_x) \geq 170 \Rightarrow N_x \geq 17/19$. But since both N_x and N_y are positive integers, we conclude that $N_x = 1, N_y = 5$; in total 6 buses.

_____

5. (a) There are two cases

$$y = |x - 1| = \begin{cases} x - 1 & \text{if } x \geq 1 \\ 1 - x & \text{if } x < 1 \end{cases}$$

For the case $y = x - 1$, substitution into the other equation gives $(x-1)^2 = 2x(x-1) + 3x + 2$. We have a quadratic equation $x^2 + 3x + 1 = 0$ with roots $x = (-3 \pm \sqrt{5})/2$ that are negative. This is an invalid solution since we assumed $x \geq 1$ at the start.

For the second case $y = 1 - x$, we get $3x^2 - 7x - 1 = 0$ which gives the valid solution $x = (7 - \sqrt{61})/6$. (The other root is ignored

since $x < 1$). The corresponding y value is $y = (\sqrt{61} - 1)/6$.

(b) Re-write the first equality as $x^2 = 2y^2 + 3$ and substitute this into the second equation,

$$\begin{aligned} y &= 9 - 2(2y^2 + 3) \\ 0 &= 4y^2 + y - 3 \\ y &= \frac{-1 \pm \sqrt{1 - 4(4)(-3)}}{2(4)} = -1 \, , \end{aligned}$$

where we ignored the other solution since $y < 0$. Thus, $x = -\sqrt{2(-1)^2 + 3} = -\sqrt{5}$.

(c) Substitute $y = 4/x$ into the other equation to get

$$\begin{aligned} x + \frac{4}{x} &= 2x \cdot \frac{4}{x} + 3 \\ \Rightarrow 0 &= x^2 - 11x + 4 \\ x &= \frac{11 \pm \sqrt{105}}{2}. \end{aligned}$$

The corresponding y is $y = \frac{11 \mp \sqrt{105}}{2}$.

(d) The method of solution is as in part (a). Consider separately the cases $3 \geq x$ and $x > 3$. The solutions are $(x, y) = (0, 3)$ and $(1, 2)$.

_____

7. (a) The roots of $y = 2x^2 - 3x - 4 = 0$ are $x_\pm = \frac{3 \pm \sqrt{41}}{4}$. The domain of x that makes $y < 0$ lies between the roots, that is

$$y < 0 \Rightarrow \frac{3 - \sqrt{41}}{4} < x < \frac{3 + \sqrt{41}}{4}.$$

Meanwhile, $2 - x < 0$ implies $x > 2$. Combining both conditions, we get $2 < x < \frac{3 + \sqrt{41}}{4}$.

(b) Solving $x^2 - x - 1 = 4$, we obtain $x_\pm = \frac{1 \pm \sqrt{21}}{2}$. Thus, $x^2 - x - 1 > 4$ requires that $x < \frac{1 - \sqrt{21}}{2}$ or $x > \frac{1 + \sqrt{21}}{2}$. Similarly, solving $3x^2 - 4 = x - 2$ gives $x = 1$ or $x = -2/3$. So, $3x^2 - 4 > x - 2$ requires that $x < -2/3$ or $x > 1$. The common domain satisfying both conditions is $x < \frac{1 - \sqrt{21}}{2}$ or $x > \frac{1 + \sqrt{21}}{2}$.

(c) The two simultaneous conditions are $2x^2 - 3x + 3 > 2 - x$ and $2 - x > 3x - 4$. The first condition is true for all x. The second condition gives $x < 3/2$.

_____

9. Suppose the rectangle has sides a and b. The perimeter is $P = 2(a + b)$ and $P^2 = 24$. In other words, $a + b = \sqrt{6}$. Also, the total area is $ab = 1$, implying that $b = 1/a$; substituting this into the first condition yields the quadratic equation $a^2 - \sqrt{6}a + 1 = 0$ whose solution is $a = \frac{\sqrt{6}\pm\sqrt{2}}{2}$ and $b = \frac{\sqrt{6}\mp\sqrt{2}}{2}$.

11. (a) The matrix is

$$X = \begin{pmatrix} 1 & 2 & 3 \\ 2 & 3 & 1 \\ 3 & 1 & 2 \end{pmatrix}.$$

The columns represent the drinks while the rows represent the ingredients.

(b) The cost matrix is

$$Y = \begin{pmatrix} 0.1 & 0.2 & 0.3 \\ 0.15 & 0.15 & 0.25 \end{pmatrix}.$$

To obtain the cost of drinks made by using materials wholly from supplier A or B, we just need to multiply Y and X,

$$YX = \begin{pmatrix} 0.1 & 0.2 & 0.3 \\ 0.15 & 0.15 & 0.25 \end{pmatrix} \cdot \begin{pmatrix} 1 & 2 & 3 \\ 2 & 3 & 1 \\ 3 & 1 & 2 \end{pmatrix}$$
$$= \begin{pmatrix} 1.4 & 1.1 & 1.1 \\ 1.2 & 1 & 1.1 \end{pmatrix}.$$

Therefore the cost table for the drinks is

	Cool	Wow	Zing
Supplier A	1.4	1.1	1.1
Supplier B	1.2	1.0	1.1

(c) Since the sale price is double the cost price, the profit matrix at each store for each drink is $P = \begin{pmatrix} 1.4 & 1.1 & 1.1 \\ 1.2 & 1.0 & 1.1 \end{pmatrix}$. At Apex, the total profit equals the first row of the profit matrix multiplied by the number of glasses sold, which is

$$\begin{pmatrix} 1.4 & 1.1 & 1.1 \end{pmatrix} \cdot \begin{pmatrix} 5 \\ 10 \\ 20 \end{pmatrix} = 40 \text{ dollars.}$$

Similarly, at Bravo the total profit is

$$\begin{pmatrix} 1.2 & 1.0 & 1.1 \end{pmatrix} \cdot \begin{pmatrix} 10 \\ 15 \\ 15 \end{pmatrix} = 43.5 \text{ dollars.}$$

Did You Know?

Try this. Generate a sequence as follows:

Start with any positive integer n.
(1) If it is even, divide it by two; that is $n \to n/2$.
(2) If it is odd, multiply by three and add one, that is $n \to 3n + 1$.

Repeat those two steps on each result. You will eventually end up with 1 and then enter the cycle $1 \to 4 \to 2 \to 1$. Try it on any natural number!

Will you always reach 1? The Collatz Conjecture, which is unproven, suggests so.

The book *Integrated Mathematics for Explorers* by W.K. Ng and R. Parwani contains not only school-level exercises, but also challenges, stimulating puzzles, and a list of unsolved mathematics problems such as the Collatz Conjecture.

View sample pages at www.simplicitysg.net/books/imaths.

Chapter 4

Trigonometry

4.1 Solutions to Exercises

1. See the figure below.

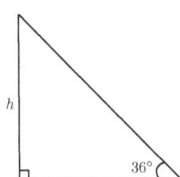

Figure 4.1: Figure for Exercise 1.

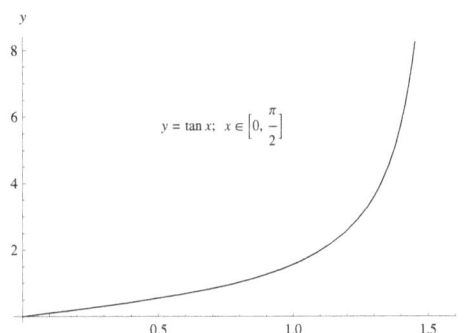

Figure 4.2: Plot of $y = \tan x$. Note that the function is not linear.

(a) To find the height of the building, take the tangent of the angle indicated in the figure below,

$$\tan 36° = \frac{h}{30}$$
$$\Rightarrow h = 30 \tan 36° = 21.8 \text{ m.}$$

(b) Let the new angle subtended to point P be α,

$$\tan \alpha = \frac{h/2}{30}$$
$$= \frac{21.8}{2(30)} = 0.363$$
$$\Rightarrow \alpha = \tan^{-1} 0.363 = 19.96°.$$

(c) Let Q be at a height h' above ground. Then

$$\tan 18° = \frac{h'}{30}$$
$$\Rightarrow h' = 30 \tan 18° = 9.75 \text{ m.}$$

(d) The locations of P and Q are different since the tangent function is not linear, for example, $\tan 2x \neq 2 \tan x$. In fact, $\tan 2x = \frac{2 \tan x}{1 - \tan^2 x}$. The graph below illustrates the non-linear behaviour of the tangent function.

2. (a) Snell's Law is

$$n_1 \sin \theta_1 = n_2 \sin \theta_2$$
$$\Rightarrow \frac{\sin \theta_2}{\sin \theta_1} = \frac{n_1}{n_2}.$$

Recall the property of $\sin \theta$: It is an increasing function for $\theta \in [0, \pi/2]$; see the figure below.

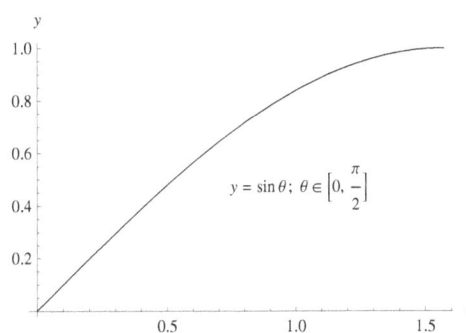

Figure 4.3: Plot of $y = \sin \theta$ for $\theta \in [0, \pi/2]$.

Therefore $\theta_2 > \theta_1 \Rightarrow \sin \theta_2 > \sin \theta_1$, and we have

$$\frac{n_1}{n_2} = \frac{\sin \theta_2}{\sin \theta_1} > 1$$
$$\Rightarrow \frac{n_1}{n_2} > 1$$
$$\therefore n_1 > n_2.$$

(b) The critical angle occurs when $\theta_2 = \pi/2$,

$$
\begin{aligned}
n_1 \sin \theta_c &= n_2 \sin \frac{\pi}{2} \\
\sin \theta_c &= \frac{n_2}{n_1} \\
\therefore \theta_c &= \sin^{-1} \frac{n_2}{n_1}.
\end{aligned}
$$

(c) [Optional]: When $\theta_1 > \theta_c$ we have total internal refraction. This is the basic principle behind optical fibres that are important in telecommunications.

---

3. (a) We can use the equation of horizontal motion $x(t) = Ut \cos \theta$ to express time t as $t = \frac{x}{U \cos \theta}$ and substitute into $y(t)$,

$$
\begin{aligned}
y(x) &= U \left(\frac{x}{U \cos \theta} \right) \sin \theta - \frac{g}{2} \left(\frac{x}{U \cos \theta} \right)^2 \\
&= x \tan \theta - \frac{g x^2}{2 U^2 \cos^2 \theta} \\
&= x \tan \theta - \frac{g x^2 \sec^2 \theta}{2 U^2}.
\end{aligned}
$$

(b) When the particle hits the ground again, $y(x) = 0$. So we have

$$
\begin{aligned}
0 &= y \big|_{x=R} \\
0 &= R \tan \theta - \frac{g R^2 \sec^2 \theta}{2 U^2} \\
&= R \left[\tan \theta - \frac{g R \sec^2 \theta}{2 U^2} \right] \\
\Rightarrow 0 = R \quad &\text{or} \quad 0 = \tan \theta - \frac{g R \sec^2 \theta}{2 U^2}.
\end{aligned}
$$

Since $R \neq 0$, we have to choose the second solution,

$$
\begin{aligned}
0 &= \tan \theta - \frac{g R \sec^2 \theta}{2 U^2} \\
\tan \theta &= \frac{g R \sec^2 \theta}{2 U^2} \\
\Rightarrow R &= \frac{2 U^2 \tan \theta}{g \sec^2 \theta} \\
&= \frac{2 U^2 \frac{\sin \theta}{\cos \theta}}{g} \times \cos^2 \theta \\
&= \frac{2 U^2 \sin \theta \cos \theta}{g} \\
&= \frac{U^2 \sin 2\theta}{g}.
\end{aligned}
$$

(c) From $R = \frac{U^2 \sin 2\theta}{g}$ we see that the range is maximum when $\sin 2\theta$ is maximum (keeping other parameters fixed). Since the maximum value of the sine function is 1, this happens when $2\theta = \frac{\pi}{2} \Rightarrow \theta = \frac{\pi}{4}$. Hence the maximum range is $R_{\max} = U^2/g$ and it is achieved when at $\theta = \frac{\pi}{4}$.

---

4. Given that $V(t) = V_0 \sin(2\pi f t)$ where $f = 60 \text{ s}^{-1}$ and $V_0 = 240\sqrt{2}$ volts.

(a) For $0 < t < \frac{1}{2f} = \frac{1}{120} \text{s}$, we have the figure below.

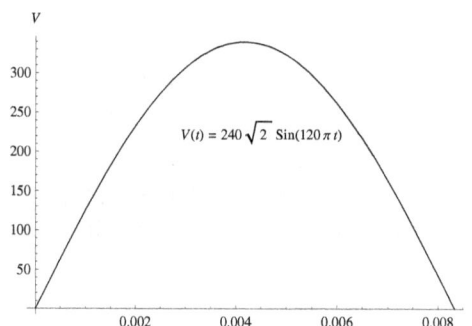

$$V(t) = 240\sqrt{2} \; \text{Sin}(120 \pi t)$$

Figure 4.4: Plot of $V(t) = V_0 \sin(2\pi f t) = 240\sqrt{2} \sin(120\pi t)$ for $t \in [0, 1/120]$.

(b) From the next figure we see that voltage is larger than $V = 240$ volts in between (t_1, t_2).

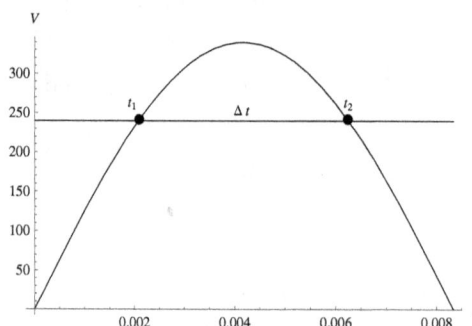

Figure 4.5: Figure to find out $t_{1,2}$ for Exercise 4(b).

We need to determine both time t_1 and t_2,

$$V(t) = 240 \Rightarrow 240\sqrt{2}\sin(2\pi ft) = 240$$

$$\sin(2\pi ft) = \frac{1}{\sqrt{2}}$$

$$2\pi ft = \sin^{-1}\frac{1}{\sqrt{2}} = \frac{\pi}{4} \text{ or } \frac{3\pi}{4}$$

$$t = \frac{\pi}{4}\times\frac{1}{2\pi f} \text{ or } \frac{3\pi}{4}\times\frac{1}{2\pi f}$$

$$= \frac{1}{8f} \text{ or } \frac{3}{8f}$$

$$\therefore t_{1,2} = \frac{1}{480} \text{ s and } \frac{1}{160} \text{ s.}$$

Hence in the interval $\frac{1}{160}$s $> t > \frac{1}{480}$s the voltage is larger than 240 volts.

(c) From the figure below, we see that voltage is less than $V = 120$ volts and decreasing from t_3 onwards until $t = \frac{1}{2f} = 0.0083$ s.

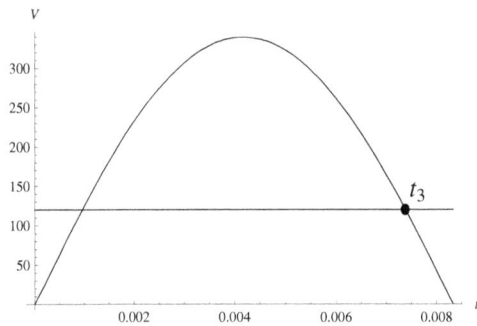

Figure 4.6: Figure to find out t_3 for Exercise 4(c).

We need to determine the time t_3,

$$V(t) = 120 \Rightarrow 240\sqrt{2}\sin(2\pi ft_3) = 120$$

$$\sin(2\pi ft_3) = \frac{1}{2\sqrt{2}}$$

$$2\pi ft_3 = \sin^{-1}\frac{1}{2\sqrt{2}} = 0.8849\pi$$

$$t_3 = 0.8849\pi\times\frac{1}{2\pi f}$$

$$\therefore t_3 = 0.0074 \text{ s.}$$

Hence when 0.0083 s $> t > 0.0074$ s the voltage is less than 120 volts and decreasing.

5. (a) For $y(L) = 0$,

$$y(L) = A\sin\frac{2\pi L}{\lambda} = 0$$

$$\Rightarrow \frac{2\pi L}{\lambda} = n\pi; \quad n = 1,2,3...$$

$$\lambda = \frac{2L}{n}.$$

6. (a)

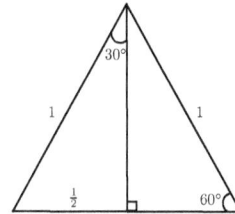

Figure 4.7: Equilateral triangle of side 1.

From the figure, we have

$$\sin 30° = \frac{\text{Opposite side}}{\text{Hypotenuse}}$$
$$= \frac{1/2}{1} = \frac{1}{2}.$$

$$\cos 60° = \frac{\text{Adjacent side}}{\text{Hypotenuse}}$$
$$= \frac{1/2}{1} = \frac{1}{2} = \sin 30°.$$

Also,

$$\sin 60° = \frac{\sqrt{1-(1/2)^2}}{1} = \frac{\sqrt{3}}{2}.$$

$$\cos 30° = \frac{\sqrt{3}/2}{1} = \frac{\sqrt{3}}{2} = \sin 60°.$$

(b)

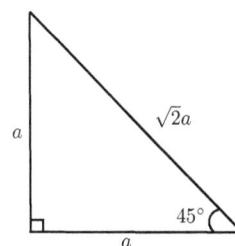

Figure 4.8: Right-angled isosceles triangle.

From the figure, we have

$$\sin 45° = \frac{a}{\sqrt{2}a} = \frac{1}{\sqrt{2}} .$$

$$\cos 45° = \frac{a}{\sqrt{2}a} = \frac{1}{\sqrt{2}} = \sin 45° .$$

(c)

$$\tan 30° = \frac{\sin 30°}{\cos 30°} = \frac{1/2}{\sqrt{3}/2} = \frac{1}{\sqrt{3}} .$$

$$\tan 45° = \frac{\sin 45°}{\cos 45°} = \frac{1/\sqrt{2}}{1/\sqrt{2}} = 1 .$$

$$\tan 60° = \frac{\sin 60°}{\cos 60°} = \frac{\sqrt{3}/2}{1/2} = \sqrt{3} .$$

7. *Note: $\sin\theta$ is an increasing function and also $1 \geq \sin\theta \geq 0$ for $\theta \in [0, \pi/2]$.*
(a) To arrange the sequence in order, we compare the other terms to $\sin 35°$. We have

$$\cos 35° = \sin(90° - 35°) = \sin 55° > \sin 35° .$$

$$\tan 55° = \frac{\sin 55°}{\cos 55°}$$

$$= \frac{\sin 55°}{\sin 35°} > 1. \qquad (4.1)$$

Since $1 \geq \sin\theta \geq 0$ for $\theta \in [0, \pi/2]$, we have $\sin 35° < \cos 35° < \tan 55°$.

(b)

$$\sec 15° = \frac{1}{\cos 15°} = \frac{1}{\sin(90° - 15°)} = \frac{1}{\sin 75°} .$$

$$\csc 15° = \frac{1}{\sin 15°} .$$

$$\frac{\sec 15°}{\csc 15°} = \frac{1/\sin 75°}{1/\sin 15°} = \frac{\sin 15°}{\sin 75°} < 1 .$$

So we have $\sec 15° < \csc 15°$.

---

8. (a)

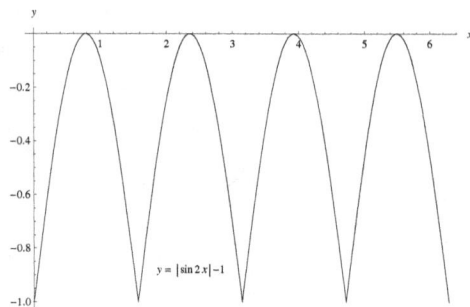

Figure 4.9: Plot of $y(x) = |\sin 2x| - 1$.

From the figure, we have the range of $y(x)$ for $x \in [0, 2\pi]$:
$-1 \leq y \leq 0$.

(b)

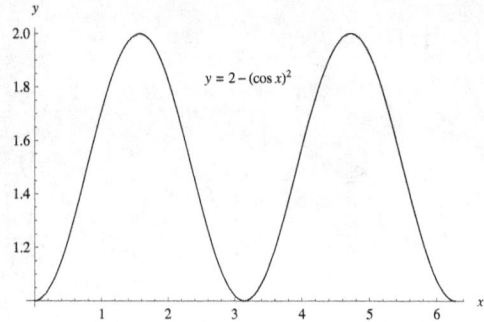

Figure 4.10: Plot of $y(x) = 2 - \cos^2 x$.

We have the range of $y(x)$ for $x \in [0, 2\pi]$:
$1 \leq y \leq 2$.

(c)

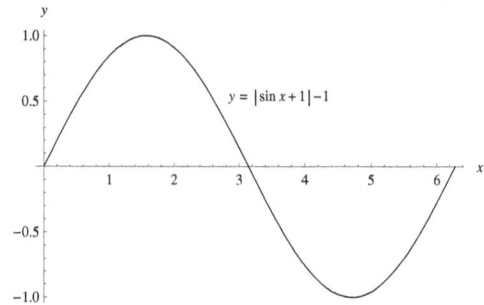

Figure 4.11: Plot of $y(x) = |(\sin x) + 1| - 1$.

We have the range of $y(x)$ for $x \in [0, 2\pi]$: $-1 \leq y \leq 1$.

(d)

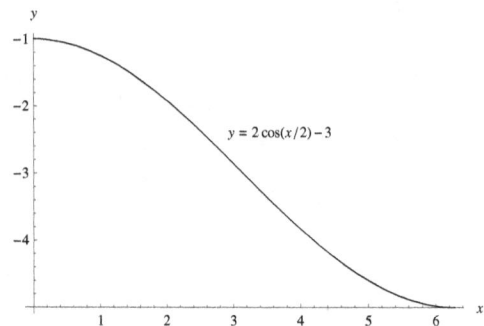

Figure 4.12: Plot of $y(x) = 2\cos(x/2) - 3$.

We have the range of $y(x)$ for $x \in [0, 2\pi]$:
$-5 \le y \le -1$.

_____

9. Given $\sin x = 1/5$ and $0 < x < \pi/2$.

(a)

$$\cos x = \sqrt{1 - \sin^2 x}$$
$$= \frac{\sqrt{24}}{5} = \frac{2\sqrt{6}}{5} = 0.4\sqrt{6}$$
$$\tan x = \frac{\sin x}{\cos x}$$
$$= \frac{1}{\sqrt{24}} = \frac{1}{2\sqrt{6}}.$$

(b)

$$\sin 2x = 2 \sin x \cos x$$
$$= 2 \times \frac{1}{5} \times \frac{2\sqrt{6}}{5}$$
$$= \frac{4\sqrt{6}}{25} ;$$
$$\cos 2x = 2 \cos^2 x - 1$$
$$= 2 \left(\frac{2\sqrt{6}}{5} \right)^2 - 1$$
$$= \frac{48}{25} - 1 = \frac{23}{25} ;$$
$$\tan 2x = \frac{2 \tan x}{1 - \tan^2 x}$$
$$= \frac{2\left(\frac{1}{2\sqrt{6}}\right)}{1 - \left(\frac{1}{2\sqrt{6}}\right)^2}$$
$$= \frac{1}{\sqrt{6}} \times \frac{24}{23}$$
$$= \frac{24}{23\sqrt{6}} \times \frac{\sqrt{6}}{\sqrt{6}} = \frac{4\sqrt{6}}{23}.$$

(Note: The same result for $\tan 2x$ follows from the ratio $\sin 2x / \cos 2x$.)

(c) *Note: Use the identities,*
$\sin(A + B) = \sin A \cos B + \cos A \sin B$,
$\cos(A + B) = \cos A \cos B - \sin A \sin B$,

and $\tan(A + B) = \frac{\tan A + \tan B}{1 - \tan A \tan B}$.

$$\sin\left(x + \frac{\pi}{3}\right) = \sin x \, \cos \frac{\pi}{3} + \cos x \, \sin \frac{\pi}{3}$$
$$= \frac{1}{5} \times \frac{1}{2} + \frac{2\sqrt{6}}{5} \times \frac{\sqrt{3}}{2}$$
$$= \frac{1 + 2\sqrt{18}}{10} = \frac{1 + 6\sqrt{2}}{10}$$
$$= 0.1 + 0.6\sqrt{2} .$$
$$\cos\left(2x + \frac{\pi}{4}\right) = \cos 2x \cos \frac{\pi}{4} - \sin 2x \sin \frac{\pi}{4}$$
$$= \frac{23}{25} \times \frac{\sqrt{2}}{2} - \frac{4\sqrt{6}}{25} \times \frac{\sqrt{2}}{2}$$
$$= \frac{23\sqrt{2} - 4\sqrt{12}}{50}$$
$$= \frac{23\sqrt{2} - 8\sqrt{3}}{50}$$
$$= 0.46\sqrt{2} - 0.16\sqrt{3} .$$

and

$$\tan\left(x + \frac{\pi}{6}\right) = \frac{\tan x + \tan \frac{\pi}{6}}{1 - \tan x \tan \frac{\pi}{6}}$$
$$= \frac{\frac{1}{2\sqrt{6}} + \frac{1}{\sqrt{3}}}{1 - \frac{1}{2\sqrt{6}} \times \frac{1}{\sqrt{3}}}$$
$$= \frac{\sqrt{3} + 2\sqrt{6}}{2\sqrt{3 \times 6} - 1}$$
$$= \frac{\sqrt{3} + 2\sqrt{6}}{6\sqrt{2} - 1} .$$

_____

10. (a)

$$1 + \sin x = 3 \cos^2 x$$
$$= 3(1 - \sin^2 x)$$
$$\Rightarrow 0 = 3 \sin^2 x + \sin x - 2$$
$$\sin x = \frac{-1 \pm \sqrt{1 - 4(3)(-2)}}{2(3)}$$
$$= \frac{-1 \pm 5}{6} = 2/3 \text{ or } -1.$$

When $\sin x = 2/3$, we have the angle $\sin^{-1}(2/3) = 0.73$ rad. On the other hand, when $\sin x = -1 \Rightarrow x = 3\pi/2 = 4.71$ rad. So the smallest positive x is $x = 0.73$ rad.

28

(b)

$$\tan^2 x = 2 - 3\sec x$$
$$\sec^2 x - 1 = 2 - 3\sec x$$
$$\Rightarrow 0 = \sec^2 x + 3\sec x - 3$$
$$\sec x = \frac{-3 \pm \sqrt{9 - 4(-3)}}{2}$$
$$= \frac{-3 \pm \sqrt{21}}{2}$$
$$= 0.791 \text{ or } -3.791.$$

Since $|\sec x| > 1$ we need to ignore the first solution $\sec x = 0.791$. When $\sec x = -3.791 \Rightarrow \cos x = -0.264$. Now, $\cos^{-1}(0.264) = 1.304$ rad. As $\cos x$ is negative, we have to consider the second or third quadrants. The smallest positive x is $x = \pi - 1.304$ rad $= 1.84$ rad.

(c)

$$\sin(x° - 20°) = \cos 80°$$
$$= \sin(90° - 80°) \text{ or } \sin[180° - (90° - 80°)]$$
$$= \sin 10° \text{ or } \sin 170°$$
$$\therefore x° - 20° = 10° \text{ or } 170°.$$

Hence the smallest positive x is $x = 30$.

(d)

$$\cos(1 - 2x) = 0.75$$
$$(1 - 2x) = \cos^{-1}(0.75)$$
$$= 0.723 \text{ rad or } (2\pi - 0.723) \text{ rad}$$
$$x = \frac{1 - 0.723}{2}, \frac{1 - (2\pi - 0.723)}{2}$$
$$= 0.14 \text{ rad or } -2.28 \text{ rad.}$$

Hence the smallest positive x is $x = 0.14$ rad.

___

11. *Note: The symmetries of trigonometric functions are:* $\sin(-A) = -\sin A$, $\cos(-A) = \cos A$, *and* $\tan(-A) = -\tan A$. *This is equivalent to saying that both the sine and tangent functions are odd while the cosine function is even in its argument.*

Given that $f(x) = x\cos x + a\sin x + b\tan x + 5$ and $f(3) = 6$. Firstly we observe that

$$f(-x) = -x\cos(-x) + a\sin(-x)$$
$$+ b\tan(-x) + 5$$
$$= -x\cos x - a\sin x - b\tan x + 5$$
$$= -[x\cos x + \sin x + b\tan x + 5] + 10$$
$$= -f(x) + 10.$$

Hence $f(-3) = -f(3) + 10 = -6 + 10 = 4$. Also, $f(0) = -f(0) + 10 \Rightarrow 2f(0) = 10 \Rightarrow f(0) = 5$.

___

12. *Note:* $a^2 - b^2 = (a - b)(a + b)$

$$\frac{1}{4} = \sin^4 x - \cos^4 x$$
$$= (\sin^2 x + \cos^2 x) \times (\sin^2 x - \cos^2 x)$$
$$= (1) \times (\sin^2 x - (1 - \sin^2 x))$$
$$= 2\sin^2 x - 1,$$

which leads to

$$\sin^2 x = \frac{1/4 + 1}{2} = \frac{5}{8}$$
$$\Rightarrow \sin x = \pm\sqrt{\frac{5}{8}}.$$

Since $x \in [0, \pi/2]$, we conclude that $\sin x = \frac{1}{2}\sqrt{\frac{5}{2}}$. The cosine and tangent are

$$\cos x = \sqrt{1 - \sin^2 x}$$
$$= \sqrt{1 - \left(\frac{1}{2}\sqrt{\frac{5}{2}}\right)^2}$$
$$= \sqrt{\frac{3}{8}} = \frac{1}{2}\sqrt{\frac{3}{2}};$$
$$\tan x = \frac{\sin x}{\cos x} = \frac{\sqrt{5}/(2\sqrt{2})}{\sqrt{3}/(2\sqrt{2})}$$
$$= \sqrt{\frac{5}{3}}.$$

___

13. *Note: We have to use both the sine and cosine rules.*

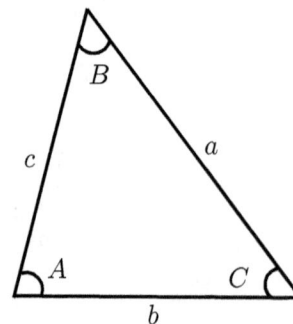

Figure 4.13: Figure of triangle for Exercise 13.

___

(a) With $a = 5, b = 7, C = 80°$, using the cosine rule gives,

$$
\begin{aligned}
c^2 &= a^2 + b^2 - 2ab\cos C \\
c &= \sqrt{5^2 + 7^2 - 2(5)(7)\cos 80°} \\
&= \sqrt{74 - 12.16} = 7.864. \\
&\approx 7.86 . \tag{4.2}
\end{aligned}
$$

Then with the sine rule,

$$
\begin{aligned}
\frac{5}{\sin A} &= \frac{c}{\sin C} \\
\sin A &= \frac{5\sin C}{c} = \frac{5\sin 80°}{7.864} \\
\Rightarrow A &= \sin^{-1} 0.626 = 38.8°; \\
\frac{7}{\sin B} &= \frac{c}{\sin C} \\
\sin B &= \frac{7\sin C}{c} = \frac{7\sin 80°}{7.864} \\
\Rightarrow B &= \sin^{-1} 0.877 = 61.2°.
\end{aligned}
$$

(b) Given $a = 6, b = 7, A = 50°$,

$$
\begin{aligned}
\frac{7}{\sin B} &= \frac{a}{\sin A} \\
\sin B &= \frac{7\sin A}{a} = \frac{7\sin 50°}{6} \\
\Rightarrow B &= \sin^{-1} 0.8937 = 63.3° \text{ or } 116.7°.
\end{aligned}
$$

We can now obtain angle C since we know both A and B. $C = 180° - (A+B) = 180° - (50° + 63.3°) = 66.7°$ or $C = 180° - (50° + 116.7°) = 13.3°$.

The length c can be determined by

$$
\begin{aligned}
c &= \sqrt{a^2 + b^2 - 2ab\cos C} \\
&= \sqrt{6^2 + 7^2 - 2(6)(7)\cos C} \\
&= \sqrt{85 - 84\cos 66.7°} \text{ or } \sqrt{85 - 84\cos 13.3°} \\
&= 7.2 \text{ or } 1.8.
\end{aligned}
$$

Hence, one solution is $B = 63.3°, C = 66.7°, c = 7.2$, while the other is $B = 116.7°, C = 13.3°, c = 1.8$.

(c) Given that $A = 50°, B = 120°, c = 7$. We have $C = 180° - (50° + 120°) = 10°$ and

$$
\begin{aligned}
\frac{a}{\sin A} &= \frac{c}{\sin C} \\
\Rightarrow a &= \frac{7}{\sin 10°} \times \sin 50° = 30.88 \approx 30.9. \\
\frac{b}{\sin B} &= \frac{c}{\sin C} \\
\Rightarrow b &= \frac{7}{\sin 10°} \times \sin 120° = 34.91 \approx 34.9.
\end{aligned}
$$

(d) Given that $A = 50°, B = 120°, a = 10$. As in part (c), we have $C = 180° - (50° + 120°) = 10°$ and

$$
\begin{aligned}
\frac{c}{\sin C} &= \frac{a}{\sin A} \\
\Rightarrow c &= \frac{10}{\sin 50°} \times \sin 10° = 2.27 \approx 2.3. \\
\frac{b}{\sin B} &= \frac{a}{\sin A} \\
\Rightarrow b &= \frac{10}{\sin 50°} \times \sin 120° = 11.31 \approx 11.3.
\end{aligned}
$$

---

14. *Note: The area of triangle ABC is given by $A = \frac{1}{2}ab\sin C$.*

(a)

$$
\begin{aligned}
\text{Area} &= \frac{1}{2}ab\sin C = \frac{1}{2}(5)(7)\sin 80° \\
&= 17.2.
\end{aligned}
$$

(b)

$$
\begin{aligned}
\text{Area} &= \frac{1}{2}ab\sin C = \frac{1}{2}(6)(7)\sin 66.7° \\
&= 19.3.
\end{aligned}
$$

and for the other possibility,

$$
\begin{aligned}
\text{Area} &= \frac{1}{2}ab\sin C = \frac{1}{2}(6)(7)\sin 13.3° \\
&= 4.8.
\end{aligned}
$$

(c)

$$
\begin{aligned}
\text{Area} &= \frac{1}{2}ab\sin C = \frac{1}{2}(30.88)(34.91)\sin 10° \\
&= 93.6.
\end{aligned}
$$

(d)

$$
\begin{aligned}
\text{Area} &= \frac{1}{2}ab\sin C = \frac{1}{2}(10)(11.31)\sin 10° \\
&= 9.8.
\end{aligned}
$$

---

15. (a) The regular pentagon inscribed in a circle of unit length is illustrated below

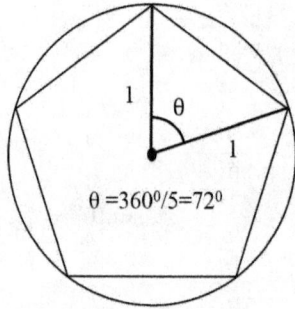

Figure 4.14: Figure of regular pentagon inscribed in a unit circle; for Exercise 15(a).

The polygon's area is

$$
\begin{aligned}
\text{Area} &= 5\frac{1}{2}ab\sin C \\
&= \frac{5}{2}(1)(1)\sin 72° \\
&= 2.38.
\end{aligned}
$$

(b) The regular pentagon circumscribes the unit circle as shown.

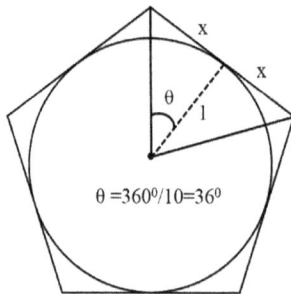

Figure 4.15: Figure of a regular pentagon circumscribing the unit circle; for Exercise 15(b).

From the figure, $\tan 36° = x$ so the side of the pentagon is $2x = 2\tan 36°$. By viewing the pentagon as combination of five triangles of base $2x$ and height 1 we have the area of pentagon,

$$
\begin{aligned}
\text{Area} &= 5\left(\frac{1}{2} \times \text{base} \times \text{height}\right) \\
&= 5\left(\frac{1}{2}(2x)(1)\right) \\
&= \frac{5}{2} \times (2\tan 36°) \\
&= 3.63.
\end{aligned}
$$

16. Consider a circle inscribed in a regular heptagon of side one,

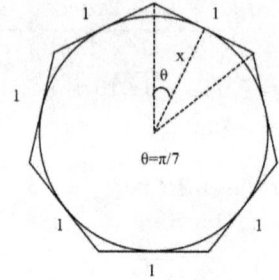

Figure 4.16: Figure of a circle inscribed in a regular heptagon of side one for Exercise 16.

From the figure,

$$
\begin{aligned}
\tan\theta &= \tan\frac{\pi}{7} = \frac{1/2}{x} \\
x &= \frac{1/2}{\tan \pi/7} = 1.04.
\end{aligned}
$$

Hence the shortest distance from the center of the circle to the heptagon is $x = 1.04$.

_____

17. *Note: A bearing denotes a direction relative to North.* Refer to the diagram below.

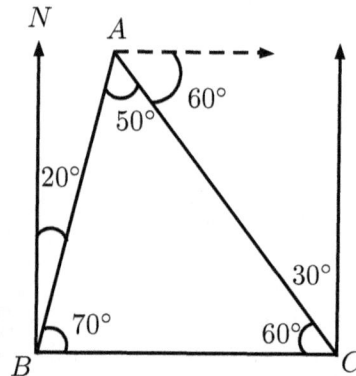

Figure 4.17: Schematic figure for Exercise 17.

(a) From the figure, the bearing of A from C is $360° - 30° = 330°$.
(b) Similarly, bearing of C from A is $90° + 60° = 150°$.
(c) Bearing of B from C is $360° - 90° = 270°$.
(d) Bearing of B from A is $90° + (60° + 50°) = 200°$.

18. Consider a triangle inscribed in a semi-circle of radius R. The area of a triangle is $\frac{1}{2}$base $\times$ height. It is easy to see (draw different triangles inscribed in the region) that the largest possible base is $2R$, and the largest corresponding height is R. So the maximum area is R^2. (We leave it to you to verify that other triangles, such as those with two vertices on the curved part and one on the diameter, will have an area less than R^2).

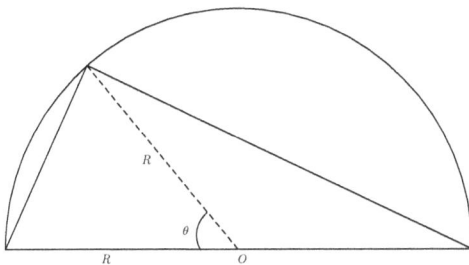

Figure 4.18: Exercise 18.

Alternatively, if you interpret the question as already fixing one side of the triangle to be the diameter (as in the figure above), then the area of that triangle is

$$
\begin{aligned}
\text{Area} &= \frac{1}{2}R^2 \sin\theta + \frac{1}{2}R^2 \sin(\pi - \theta) \\
&= \frac{R^2}{2}\left(\sin\theta + \sin\pi\cos\theta - \cos\pi\sin\theta\right) \\
&= R^2 \sin\theta.
\end{aligned}
$$

The maximum area occurs when $\sin\theta = 1$ or $\theta = \pi/2$. Hence the maximum area of the inscribed triangle is $A_{\max} = R^2$.

4.2 Solutions to Odd-Numbered Problems

1. The situation is represented in the figure below with $\angle ASP = 10°$, $\angle BSP = 15°$. So $\tan 10° = \frac{AP}{SP}$ and $\tan 15° = \frac{BP}{SP}$. Also, $AP + PB = 500$.

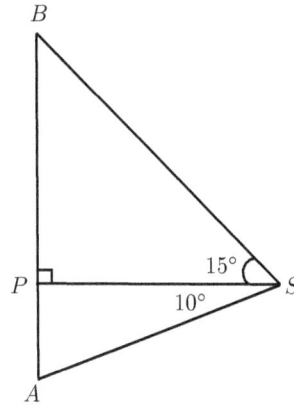

Figure 4.19: Schematic diagram for Problem 1.

Therefore

$$
\begin{aligned}
500 &= SP(\tan 10° + \tan 15°) \\
SP &= \frac{500}{(\tan 10° + \tan 15°)} \approx 1125.4\text{m}.
\end{aligned}
$$

Thus Eona is 1125.4 metres away from the shore.

3. Given $y(x) = A + B\cos(kx + C)$. Since the cosine function is bounded by $-1 \le \cos x \le 1$ for all x, and also given that the maximum (minimum) values of y are 30m (20m), we have

$$
\begin{aligned}
y_{\max} &= 30 = A + B, \\
y_{\min} &= 20 = A - B,
\end{aligned}
$$

which imply $A = 25$ and $B = 5$. The function is thus $y = 25 + 5\cos(kx + C)$. At the start of the track $x = 0, y = 25$, we have $\cos C = 0 \Rightarrow C = \pi/2$. At a peak, $y = 30$, so

$$
\begin{aligned}
30 &= 25 + 5\cos\left(kx + \frac{\pi}{2}\right) \\
1 &= \cos\left(kx + \frac{\pi}{2}\right) \\
\Rightarrow kx + \frac{\pi}{2} &= 0, 2\pi, 4\pi, \dots
\end{aligned}
$$

$$(4.3)$$

Let $\delta x = 50$ be the distance between two peaks. Then

$$
\begin{aligned}
k(\delta x) &= 2\pi \\
k &= \frac{2\pi}{(\delta x)} = \frac{2\pi}{50}.
\end{aligned}
$$

Hence $y(x) = 25 + 5\cos\left(\dfrac{2\pi x}{50} + \dfrac{\pi}{2}\right)$.

5. Given $\tan A = 3/7$, with A in the first quadrant.

(a) You can use a right-angled triangle and Pythagoras' theorem to get

$$\sin A = \frac{3}{\sqrt{3^2 + 7^2}} = \frac{3}{\sqrt{58}}.$$

(b) From the double-angle formula,

$$\cos(A/2) = \sqrt{\frac{\cos A + 1}{2}} = \sqrt{\frac{7/\sqrt{58} + 1}{2}}$$
$$= \sqrt{\frac{1}{2} + \frac{7}{2\sqrt{58}}}.$$

7. Given $\tan A = -3$, with A in the second quadrant.

(a) The double angle formula gives $\cos 2A = 2\cos^2 A - 1$. But we also know $\cos^2 A = 1/\sec^2 A = 1/(1 + \tan^2 A)$. Therefore

$$\cos 2A = 2\cos^2 A - 1$$
$$= \frac{2}{10} - 1 = -\frac{4}{5}.$$

(b) We first deduce that $\cos A = -1/\sqrt{10}$. Then

$$\sin(A/2) = \sqrt{\frac{1 - \cos A}{2}} = \sqrt{\frac{1 - (-1/\sqrt{10})}{2}}$$
$$= \sqrt{\frac{1}{2} + \frac{\sqrt{10}}{20}} = \frac{1}{10}\sqrt{50 + 5\sqrt{10}}.$$

9. Given $\sin A = 0.2$, with A in the first quadrant, and $\cos B = -0.3$, with B in the second quadrant, we have $\cos A = +\sqrt{1 - (0.2)^2} = \sqrt{0.96}$, $\tan A = \frac{0.2}{\sqrt{0.96}} \approx 0.204$; $\sin B = +\sqrt{1 - (-0.3)^2} = \sqrt{0.91}$, $\tan B = -\frac{\sqrt{0.91}}{0.3} \approx -3.180$.

(a)

$$\tan(A + B) = \frac{\tan A + \tan B}{1 - \tan A \tan B}$$
$$= \frac{0.204 - 3.180}{1 - (0.204)(-3.180)} = -1.80.$$

(b)

$$\sin B + \cos A = \sqrt{0.91} + \sqrt{0.96}$$
$$\approx 1.93.$$

11. Since the sine and cosine values are bounded between -1 and 1, the maximum and minimum values of the expression in Problem (10) are $\pm R$.

13. (a) We can treat the equation as a quadratic equation in the variable $\text{ver}(A)$:

$$0 = \text{ver}^2 A - 2\text{ver}A + \sin^2 A$$
$$\text{ver}A = \frac{2 \pm \sqrt{4 - 4(\sin A)^2}}{2}$$
$$= 1 \pm \cos A .$$

15. We can re-arrange the equation and simplify it using the factor formulae,

$$0 = 7(\cos 3x + \cos 2x) + 5(\sin 2x - \sin 3x)$$
$$= 7\left(2\cos\frac{5x}{2}\cos\frac{x}{2}\right) + 5\left(2\cos\frac{5x}{2}\sin\frac{-x}{2}\right)$$
$$= \cos\frac{5x}{2}\left(14\cos\frac{x}{2} - 10\sin\frac{x}{2}\right)$$

So, we have either $\cos\frac{5x}{2} = 0 \Rightarrow \frac{5x}{2} = \frac{\pi}{2}, \frac{3\pi}{2}, \frac{5\pi}{2}...$ or $\tan\frac{x}{2} = \frac{7}{5} \Rightarrow \frac{x}{2} = 0.9505, 4.092....$ Thus, the three smallest positive x values are $x = 0.63, 1.88, 1.90$.

17. (a) On substitution,

$$3 = g(2) = a\sin^2 2 - b\cos 2 + 2c .$$
and
$$5 = g(-2) = a\sin^2(-2) - b\cos(-2) - 2c$$
$$= a\sin^2(2) - b\cos(2) - 2c .$$

Taking the difference between the two equations, we get $4c = -2 \Rightarrow c = -\frac{1}{2}$.

(b) $g(1) = 1 = a\sin^2 1 - b\cos 1 - \frac{1}{2}$. This allows us to write $b = \frac{a\sin^2 1 - 3/2}{\cos 1}$ and thus (from part a),

$$3 = a\sin^2 2 - \left(\frac{a\sin^2 1 - 3/2}{\cos 1}\right)\cos 2 - 1$$
$$a = \frac{4 - 3\cos 2/(2\cos 1)}{\sin^2 2 - \sin^2 1 \cos 2/\cos 1}$$
$$= \frac{5.1553}{1.3722} \approx 3.76$$

Then by direct substitution,

$$b = \frac{3.76 \sin^2 1 - 3/2}{\cos 1} \approx 2.15.$$

---

19. (a) Using the R formula

$$\begin{aligned} y &= 7\sin\theta + 3\cos\theta \\ &= \sqrt{7^2 + 3^2} \, \sin\left(\theta + \tan^{-1} 3/7\right) \\ &= \sqrt{58} \, \sin\left(\theta + \tan^{-1} 3/7\right) \end{aligned}$$

Hence, $y_{\max} = \sqrt{58}$; $y_{\min} = -\sqrt{58}$.

(b)

$$\begin{aligned} y &= \cos\theta - \sin\theta \\ &= \sqrt{1^2 + (-2)^2} \, \cos\left(\theta + \tan^{-1} 1\right) \\ &= \sqrt{5} \, \cos\left(\theta + \tan^{-1} 1\right) \end{aligned}$$

Hence, $y_{\max} = \sqrt{5}$; $y_{\min} = -\sqrt{5}$.

---

21. (a) We have plotted $y = 1$ and $y = \tan x - x$ below. The intersection point is at $x = 1.1$.

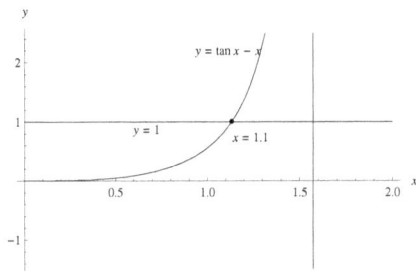

Figure 4.20: Problem 21(a).

(b) We have plotted $y = -1$ and $y = \sin x - \tan 3x$ below. The intersection point is at $x = 0.3$.

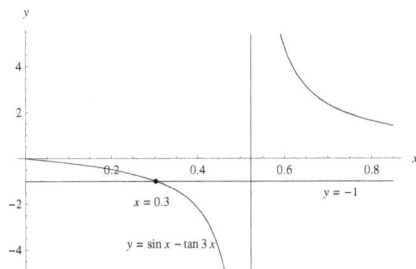

Figure 4.21: Problem 21(b).

(c) We have plotted $y = e^{-x} + \cot(x)$ and $y = 0$ below. The intersection point is at $x = 1.7$.

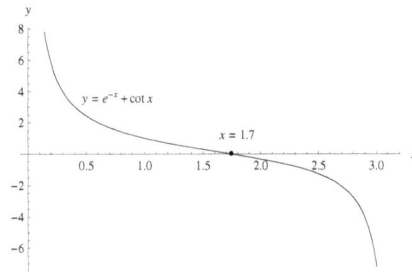

Figure 4.22: Problem 21(c).

---

23. (a) Using the factor formula, we get

$$\begin{aligned} H(t) &= 3.5 + 1.5\cos\left(\frac{2\pi t}{24}\right) \\ &\quad + 1.5\cos\left(\frac{2\pi t}{12} - 4\right) \\ &= 3.5 + 3\cos\left(\frac{\pi t}{8} - 2\right)\cos\left(\frac{\pi t}{24} - 2\right). \end{aligned}$$

Since the cosine function is bounded between -1 and $+1$, the product of cosines must be between $+1$ and -1. Thus

$$0.5 \leq H(t) \leq 6.5 .$$

(b) At $H(t) = 3.5$ m, the first condition $\cos\left(\frac{\pi t}{8} - 2\right) = 0$ implies

$$\begin{aligned} \frac{\pi t}{8} - 2 &= \pm\left[\frac{(2n-1)\pi}{2}\right] \; ; \quad n = 1, 2, 3... \\ t &= \frac{16}{\pi} \pm 4(2n-1) \\ &= \frac{16}{\pi} - 4, \frac{16}{\pi} + 4, ... \end{aligned}$$

where we only consider $t \geq 0$ solutions. The other condition is $\cos\left(\frac{\pi t}{24} - 2\right) = 0$ which implies

$$\begin{aligned} \frac{\pi t}{24} - 2 &= \pm\left[\frac{(2n-1)\pi}{2}\right] \; ; \quad n = 1, 2, 3... \\ t &= \frac{48}{\pi} \pm 12(2n-1) \\ &= \frac{48}{\pi} - 12, \frac{48}{\pi} + 12, ... \end{aligned}$$

Hence, the earliest time when the tide is at 3.5 m is when $t = \left(\frac{16}{\pi} - 4\right)$.

25. Given that $a/b = 2$, $\sin B / \sin C = 0.898$ and $c = 5.57$.

(a) From the sine rule, $a/\sin A = b/\sin B = c/\sin C \Rightarrow b = 0.898c = 5.002$. Also, $a = 2b = 2 \times 5.002 = 10.004$. From the cosine rule,

$$
\begin{aligned}
C &= \cos^{-1}\left(\frac{a^2 + b^2 - c^2}{2ab}\right) \\
&= \cos^{-1}\left(\frac{94.075}{100.080}\right) = 19.95° \approx 20°. \\
B &= \sin^{-1}\left(\frac{b}{c}\sin C\right) \\
&= \sin^{-1}(0.898 \sin 20°) \\
&= 17.89° \approx 17.9°. \\
A &= 180° - (20° + 17.9°) = 142.1°.
\end{aligned}
$$

Thus, the smallest angle in the triangle is $\angle B = 17.9°$. (Note that the smallest angle must be opposite the shortest side. So we could also have evaluated B directly).

(b) The length of the longest side is $a = 10$.

(c)

$$
\begin{aligned}
\text{Area} &= \frac{1}{2}ab\sin C \\
&= \frac{1}{2}(10)(5)\sin 19.95° = 8.5.
\end{aligned}
$$

27. (a) Consider the triangle below and let the distance from C to the side be x.

$$
\tan 30° = \frac{x}{0.5} \Rightarrow x = 0.5 \times 1/\sqrt{3} = \frac{1}{2\sqrt{3}}.
$$

So the distance from C to a side is $\frac{1}{2\sqrt{3}}$. Let the distance from C to the vertex be y,

$$
\cos 30° = \frac{0.5}{y} \Rightarrow y = \frac{0.5}{\sqrt{3}/2} = \frac{1}{\sqrt{3}}.
$$

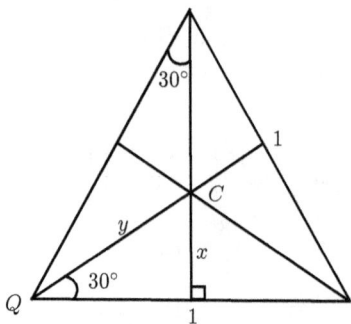

Figure 4.23: Problem 27(a).

(b) Consider a vertical slice of the pyramid like the following.

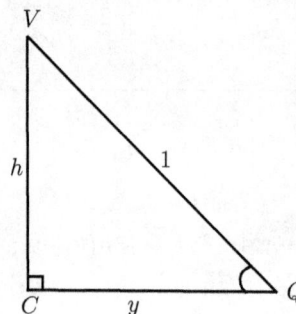

Figure 4.24: Problem 27(b).

(i) By Pythagoras rule,

$$
h = \sqrt{1^2 - y^2} = \sqrt{1 - 1/3} = \sqrt{\frac{2}{3}}.
$$

(ii) Let θ be the angle of elevation from a point P on the perimeter. Then $\tan\theta = h/r$, where r is the distance from P to the line VC of the previous figure. The largest angle of elevation happens at smallest r, that is when $r = x$, so

$$
\begin{aligned}
\theta_{max} &= \tan^{-1}\frac{h}{x} = \tan^{-1}\frac{\sqrt{2/3}}{1/(2\sqrt{3})} \\
&= \tan^{-1} 2\sqrt{2} = 70.53°.
\end{aligned}
$$

29. Let the height of the tower be h.

$$
\begin{aligned}
\tan 65° &= \frac{210 + h}{100} \\
h &= 100 \tan 65° - 210 = 4.45 \text{ m}.
\end{aligned}
$$

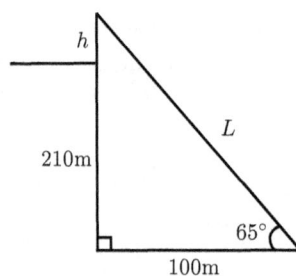

Figure 4.25: Schematic figure for Problem 29.

(b)

$$
L = \sqrt{100^2 + (210 + 4.45)^2} = 236.6 \text{ m}.
$$

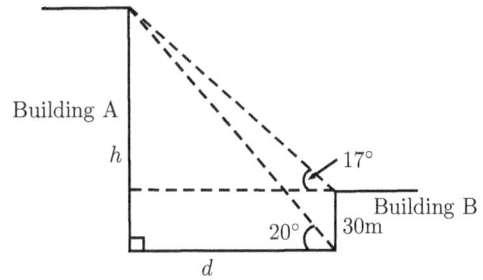

Figure 4.27: Schematic figure for Problem 33.

31. The radius is $r = \frac{1}{2}$ since the the circle is inscribed in a unit square. Let x be half of a side of the triangle. Then

$$\cos 30° = \frac{x}{r} \Rightarrow x = \frac{\sqrt{3}}{2} \times \frac{1}{2} = \frac{\sqrt{3}}{4}.$$

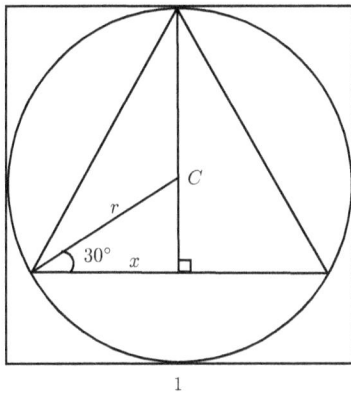

Figure 4.26: Problem 31.

Thus, the area of the triangle is

$$
\begin{aligned}
A &= \frac{1}{2}(2x)(2x)\sin 60° \\
&= \frac{3\sqrt{3}}{16}.
\end{aligned}
$$

33. Let the height of building A be h and the distance from it to building B be d (see figure).

(a)

$$
\begin{aligned}
\tan 20° &= \frac{h}{d}. \\
\tan 17° &= \frac{h-30}{d} = \frac{h-30}{h/\tan 20°}. \\
\Rightarrow h - 30 &= \frac{\tan 17°}{\tan 20°}h \\
h &= \frac{30}{1 - \tan 17°/\tan 20°} = 187.5 \text{ m.}
\end{aligned}
$$

(b)

$$d = \frac{h}{\tan 20°} = 515.1 \text{ m.}$$

35. The path taken by the scout is shown in the next figure:

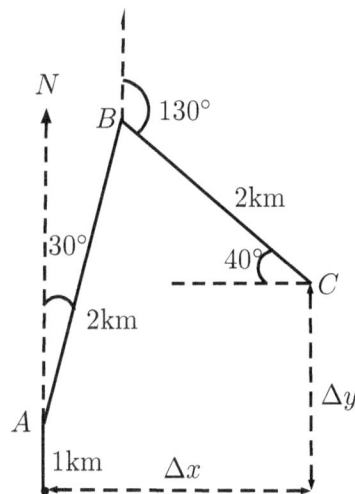

Figure 4.28: Schematic figure for Problem 35.

(a)

$$
\begin{aligned}
\Delta x &= 2\sin 30° + 2\cos 40° \\
&= 2.532 \text{ m.} \\
\Delta y &= 1 + (2\cos 30° - 2\cos 50°) \\
&= 1.446 \text{ m.} \\
D &= \sqrt{(\Delta x)^2 + (\Delta y)^2} = 2.92 \text{ m.}
\end{aligned}
$$

(b)

$$
\begin{aligned}
\text{Bearing} &= \pi + \tan^{-1}\frac{\Delta x}{\Delta y} \\
&= 240.3°.
\end{aligned}
$$

Chapter 5

Coordinate Geometry

5.1 Solutions to Exercises

1. *Recall: The mid-point of a line joining two points (x_1, y_1) and (x_2, y_2) is given by $\left(\frac{x_1+x_2}{2}, \frac{y_1+y_2}{2}\right)$. The slope of a straight line is $m = (y_2 - y_1)/(x_2 - x_1)$. The straight line equation is $y = mx + c$, where m is the slope and c the y-intercept.*

(a) Given the points $C(3, 5)$ and $D(4, 7)$. Mid-point of CD is

$$\left(\frac{x_1 + x_2}{2}, \frac{y_1 + y_2}{2}\right)$$
$$= \left(\frac{3+4}{2}, \frac{5+7}{2}\right) = \left(\frac{7}{2}, 6\right).$$

(b) The line parallel to CD has the same slope as CD,

$$m = \frac{7-5}{4-3} = 2,$$

and so the equation of the line is

$$y = 2x + 0 = 2x.$$

(c) The line perpendicular to CD has slope $-1/m = -1/2$. Also, given that the line passes through $C(3, 5)$,

$$\frac{y-5}{x-3} = -\frac{1}{2}$$
$$\Rightarrow y - 5 = -\frac{1}{2}(x-3) = -\frac{x}{2} + \frac{3}{2}$$
$$\therefore y = -\frac{x}{2} + \frac{13}{2}.$$

(d) The area of the triangle is given by

$$\text{Area}$$
$$= \frac{1}{2}\left|x_1(y_2 - y_3) + x_2(y_3 - y_1)\right.$$
$$\left. + x_3(y_1 - y_2)\right|$$
$$= \frac{1}{2}\left|0(5-7) + 3(7-0) + 4(0-5)\right|$$
$$= \frac{1}{2}|21 - 20| = \frac{1}{2}.$$

(e) Let h be the shortest distance from O to CD produced (that path will be perpendicular to CD). Therefore the area of triangle OCD will be $h|CD|/2$. Comparing with part (d), we have

$$\frac{1}{2}h|CD| = \frac{1}{2}$$
$$h\sqrt{(4-3)^2 + (7-5)^2} = 1$$
$$h = \frac{1}{\sqrt{5}}.$$

2. (a) The line $2x + 3y - 1 = 0 \Rightarrow y = \frac{-2x+1}{3}$ has slope $m = -2/3$. So a straight line that passes through the point $(2, 3)$ and is parallel to $2x + 3y - 1 = 0$ can be written as

$$\frac{y-3}{x-2} = -\frac{2}{3}$$
$$y - 3 = \frac{-2}{3}(x-2) = -\frac{2x}{3} + \frac{4}{3}$$
$$y = -\frac{2x}{3} + \frac{13}{3}.$$

(b) The normal to the line $2x + 3y - 1 = 0$ has slope $\frac{-1}{(-2/3)} = 3/2$. So the straight line that passes through the point $(2, 3)$ and is normal to $2x + 3y - 1 = 0$ is

$$\frac{y-3}{x-2} = \frac{3}{2}$$
$$y - 3 = \frac{3}{2}(x-2) = \frac{3x}{2} - 3$$
$$y = \frac{3x}{2}.$$

(c) A sketch of the lines is below.

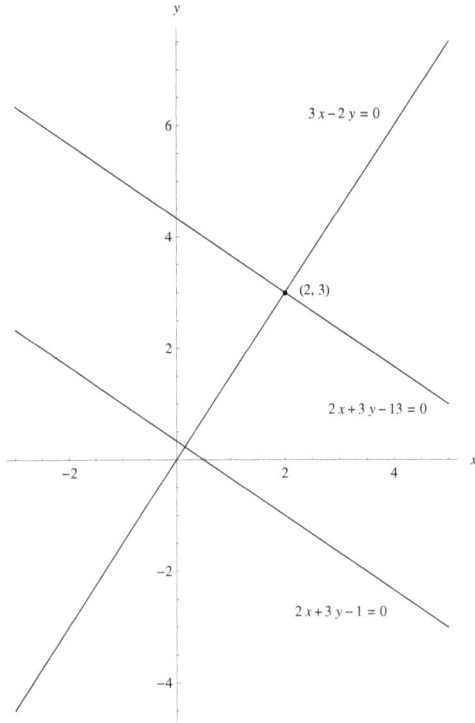

Figure 5.1: Plot of all lines in Exercise 2.

3. The points P and Q are $P(x_1, 0)$ and $Q(0, y_2)$ respectively.

(a) By substitution, at P we have $y_1 = 0$,

$$2x_1 - 1 = 0$$
$$\Rightarrow x_1 = 1/2$$
$$\therefore P = (1/2, 0).$$

Similarly, by setting $x_2 = 0$, at Q we have

$$3y_2 - 1 = 0$$
$$\Rightarrow y_2 = 1/3$$
$$\therefore Q = (0, 1/3).$$

(b) Using the mid-point formula as in Exercise (1), we have

$$R = \left(\frac{1/2 + 0}{2}, \frac{0 + 1/3}{2} \right)$$
$$= \left(\frac{1}{4}, \frac{1}{6} \right).$$

(c) The slope of PQ is $m = (1/3 - 0)/(0 - 1/2) = -2/3$. The line perpendicular to PQ

will have slope $-1/m = 3/2$. Given that the line passes through the point R, thus

$$\frac{y - 1/6}{x - 1/4} = \frac{3}{2}$$
$$2(y - 1/6) = 3(x - 1/4)$$
$$\Rightarrow 2y - 1/3 = 3x - 3/4$$
$$\therefore y = \frac{3x}{2} - \frac{5}{24}.$$

(d) Recall the points: $O(0,0)$, $P(1/2, 0)$, $Q(0, 1/3)$, $R(1/4, 1/6)$.

Area(OPR)
$$= \frac{1}{2} \left| x_1(y_2 - y_3) + x_2(y_3 - y_1) + x_3(y_1 - y_2) \right|$$
$$= \frac{1}{2} \left| 0(0 - 1/6) + 1/2(1/6 - 0) + 1/4(0 - 0) \right|$$
$$= \frac{1}{2} \left| \frac{1}{12} \right| = \frac{1}{24}.$$

Area(OQR)
$$= \frac{1}{2} \left| x_1(y_2 - y_3) + x_2(y_3 - y_1) + x_3(y_1 - y_2) \right|$$
$$= \frac{1}{2} \left| 0(\frac{1}{2} - \frac{1}{6}) + 0(\frac{1}{6} - 0) + \frac{1}{4}(0 - \frac{1}{3}) \right|$$
$$= \frac{1}{2} \left| -\frac{1}{12} \right| = \frac{1}{24}.$$

(e) The figure is shown below.

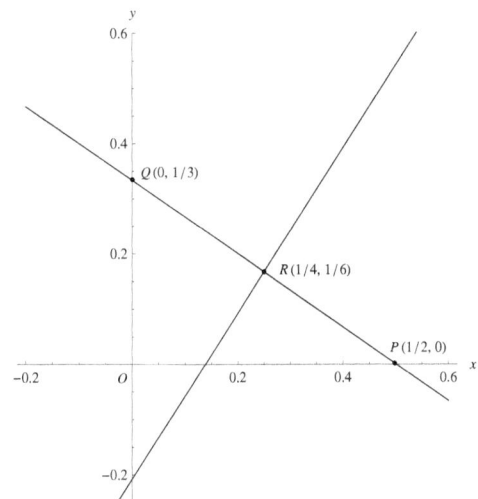

Figure 5.2: All points mentioned in Exercise 3.

4. Equation $(x-2)^2 + (y-1)^2 = 9$ represents a circle centred at $C(2,1)$ with radius of 3 as below,

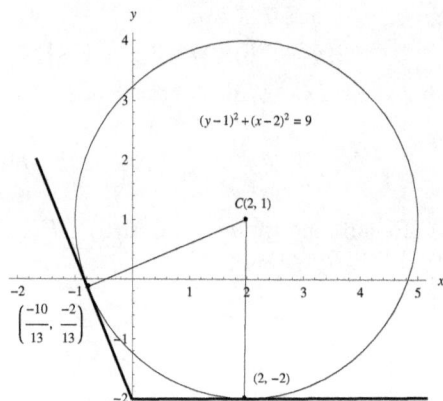

Figure 5.3: Circle $(x-2)^2 + (y-1)^2 = 9$ centred at $C(2,1)$. The bold lines represent the two possibilities for $y = mx - 2$ to meet the circle at only one point.

From Fig.(5.3) we see that if the straight line $y = mx - 2$ intercepts the circle at exactly one point, the radius pointing at that point must be perpendicular to the straight line. Thus we have the equation of normal

$$\frac{y-1}{x-2} = -\frac{1}{m},$$

where $m = (y+2)/x$. Together with the equation of the circle, we can solve the following simultaneous equations to obtain the intersection point (x,y),

$$\frac{y-1}{x-2} = -\frac{x}{y+2} \; ; \; (5.1)$$
$$(x-2)^2 + (y-1)^2 = 9.$$

The intersection happens at $y < 0$, so we can express the second equation as $y = -\sqrt{9-(x-2)^2} + 1$. We substitute this into (5.1),

$$(y-1)(y+2) = -x^2 + 2x$$
$$9 - (x-2)^2 - 3\sqrt{9-(x-2)^2} = -x^2 + 2x .$$

Expanding the left hand side and simplifying gives,

$$2x + 5 = 3\sqrt{9-(x-2)^2} .$$

We can now square both sides to solve for x,

$$9(9 - (x-2)^2) = (2x+5)^2$$
$$9(-x^2 + 4x + 5) = 4x^2 + 20x + 25$$
$$\Rightarrow 0 = 13x^2 - 16x - 20$$
$$\therefore x = \frac{16 \pm \sqrt{16^2 - 4(13)(-20)}}{2(13)}$$
$$= \frac{16 \pm \sqrt{1296}}{26}$$
$$= \frac{16 \pm 36}{26} = 2 \text{ or } -\frac{10}{13}.$$

The corresponding y values are $y = -\sqrt{9-(2-2)^2} + 1 = -2$ for $x = 2$ and $y = -\sqrt{9 - \left(-\frac{10}{13} - 2\right)^2} + 1 = -\frac{\sqrt{225}}{13} + 1 = \frac{-15+13}{13} = -\frac{2}{13}$ for $x = -\frac{10}{13}$. Thus, the intersection points are $A(2,-2)$ or $B\left(-\frac{10}{13}, -\frac{2}{13}\right)$. The solution A has slope $m = 0$ while solution B gives slope $m = -\frac{12}{5}$. So, when $m = 0$ or $m = -\frac{12}{5}$ the line $y = mx - 2$ intercepts the circle exactly once.

From the figure we see that for $-\frac{12}{5} < m < 0$, there will be no intersection points, while for $m > 0$ or $m < -\frac{12}{5}$ there will be two intersection points in each case.

(A different method of solving this problem is to use the discriminant of the quadratic equation that is obtained after combining the two simultaneous equations, see Ex.8 and 9c of Chap.2).

---

5. Let the proportionality constant be k. Then Kepler's law is

$$T^2 \propto R^3 \Rightarrow T^2 = kR^3.$$

To determine k, we can plot $Y = T^2$ against $X = R^3$, that is, $Y = kX$. The proportionality constant is given by the slope of the graph of Y against X.

On the other hand, we can take natural logarithms to obtain $2\ln T = 3\ln R + \ln k \Rightarrow \ln T = \frac{3\ln R}{2} + \frac{\ln k}{2}$. We can plot $Y = \ln T$ against $X = \ln R$. The y-intercept of this straight line will gives us the proportionality constant through $\frac{\ln k}{2}$.

---

6. (a) We plot $y = \log T$ against $x = \log a$ below.

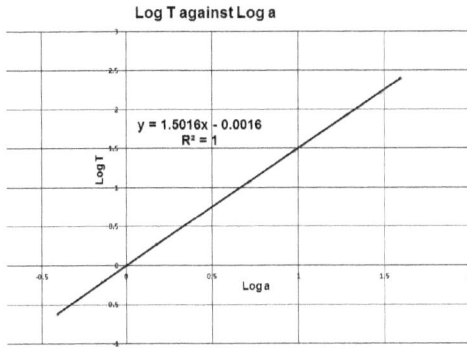

Figure 5.4: The plot of $\log T$ against $\log a$.

From the plot, the line of best fit is $Y = 1.5016X - 0.0016$ (with linear correlation coefficient R almost 1). So, we conclude that we have a straight line $\log T \approx \frac{3}{2} \log a$ (approximately).

The base of the logarithm does not matter since $\dfrac{\log_b T}{\log_b c} = \dfrac{3}{2}\dfrac{\log_b a}{\log_b c} \Rightarrow \log_c T = \dfrac{3}{2}\log_c a$.

(b) From $\log T \approx \frac{3}{2}\log a = \log a^{3/2}$, we have $T \approx a^{3/2}$.

(c) The result is consistent with Kepler's law in the previous question.

---

7. Since the variables β^2 and v^2 are expected to be linearly related, we can write

$$\beta^2 = pv^2 + q\ ,$$

where p, q are constants to be determined. They are the slope and y-intercept of the plot.

(a) Given that $(v, \beta) = (0, 1)$ and $(c, 0)$ lie on the line, so by direct substitution we have

$$
\begin{aligned}
1^2 &= p(0)^2 + q \\
\Rightarrow q &= 1 \\
\text{and } 0^2 &= pc^2 + q = pc^2 + 1 \\
\Rightarrow p &= -\frac{1}{c^2} \\
\Rightarrow \beta^2 &= 1 - \frac{v^2}{c^2} \\
\therefore \beta &= +\sqrt{1 - \frac{v^2}{c^2}}\ ,
\end{aligned}
$$

where we took the positive square-root since $\beta > 0$.

(b) If β is real, we need $1 - \frac{v^2}{c^2} \geq 0$ or equivalently $\frac{v^2}{c^2} \leq 1 \Rightarrow v \leq c$. The maximum value of v is c.

8. (a)

$$
\begin{aligned}
I &= = Ae^{-\mu d} \\
\Rightarrow \ln \frac{I}{A} &= \ln e^{-\mu d} = -\mu d.
\end{aligned}
$$

The plot of $\ln \frac{I}{A}$ versus d can be written as the straight line $Y = mX + c$, with slope m. In our context, the slope is given by $-\mu$.

(b) From the data we see that lead is the better shield because, for the same thickness, there is a bigger decrease in radiation intensity for lead compared to concrete. This conclusion is supported by the results in part (c) which show a bigger value for lead's attenuation coefficient compared to that for concrete.

(c) From part (a) and the figures below, we obtain $\mu_{\text{lead}} \approx 2.6\text{mm}^{-1}$ and $\mu_{\text{conc}} \approx 0.048\text{mm}^{-1}$.

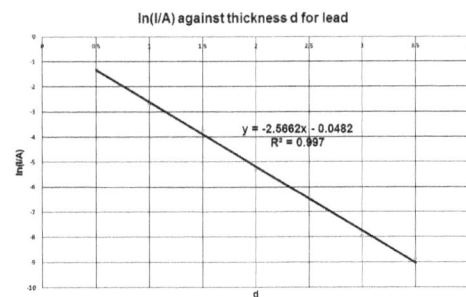

Figure 5.5: The plot of $\ln I/A$ against d for lead.

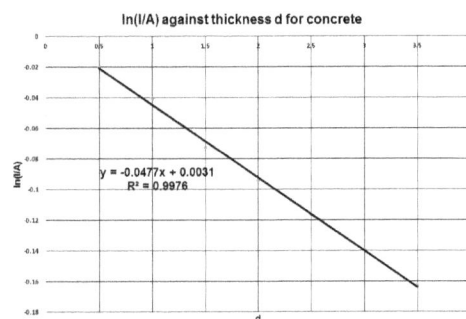

Figure 5.6: The plot of $\ln I/A$ against d for concrete.

(d) To obtain the same shielding effect we have to equate $\ln(I/A)$ for both materials. Let "1" and "2" label lead and concrete respectively.

We have

$$\begin{aligned}
\mu_1 d_1 &= \mu_2 d_2 \\
d_2 &= \frac{\mu_1}{\mu_2} \times d_1 \\
&= \frac{2.6}{0.048} \times (1 \text{ mm}) \\
&\approx 54 \text{ mm.}
\end{aligned}$$

---

9. *Recall the trigonometric identity:* $\sin^2 x + \cos^2 x = 1$. *Also, the equation of a circle is given by* $(x-a)^2 + (y-b)^2 = r^2$, *whereby the center is* (a,b) *and the radius is* r.

Given $x = A\sin\omega t$, $y = A\cos\omega t$, we can rewrite them as $\frac{x}{A} = \sin\omega t$; $\frac{y}{A} = \cos\omega t$, and use the trigonometry identity

$$\begin{aligned}
\sin^2\omega t + \cos^2\omega t &= 1 \\
\Rightarrow \frac{x^2}{A^2} + \frac{y^2}{A^2} &= 1 \\
x^2 + y^2 &= A^2.
\end{aligned}$$

Comparing this to the standard equation of a circle, we deduce that the particle motion is circular, with center at $(0,0)$ and radius A.

---

10. (a) See the figure below. Given that the largest y-coordinate is $P(3,10)$ and the y-axis is tangent to the curve C, this implies that the radius is $r = 3$ and the centre of the circle is at $(3,7)$. Thus the equation of the circle is $(x-3)^2 + (y-7)^2 = 3^2$.

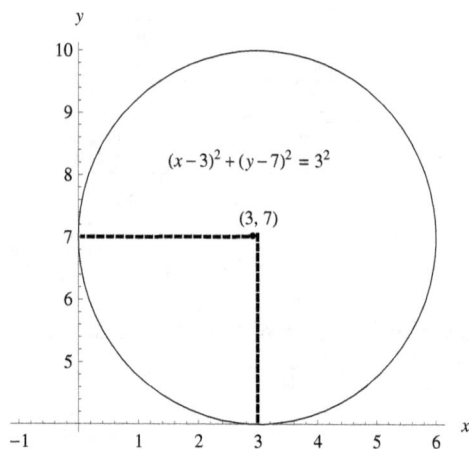

Figure 5.7: The plot of $(x-3)^2 + (y-7)^2 = 3^2$.

(b) The point with lowest y-coordinate is situated 3 units vertically below the center of the circle, at $(3,4)$.

(c) Reflection about the x-axis gives a circle centred at $(3,-7)$ with the same radius. Hence the new circle is $(x-3)^2 + (y+7)^2 = 3^2$.

(d) As in part (c), after reflection about the y-axis, the center of the new circle is at $(-3,7)$ and its equation is $(x+3)^2 + (y-7)^2 = 3^2$.

---

11. (a) The two sides of the rectangle satisfy the equations $2x + 3y - 1 = 0$ and $3x - 2y - 2 = 0$. To get the intersection point, rewrite the first equation as $y = \frac{1-2x}{3}$ and substitute into second equation,

$$\begin{aligned}
3x - 2\left(\frac{1-2x}{3}\right) &= 2 \\
3x - \frac{2}{3} + \frac{4x}{3} &= 2 \\
\frac{13x}{3} &= 2 + \frac{2}{3} = \frac{8}{3} \\
\therefore x &= \frac{8}{13} \\
\therefore y &= \frac{1 - 2(8/13)}{3} \\
&= \frac{13 - 16}{39} = -\frac{1}{13}.
\end{aligned}$$

Hence point A is $\left(\frac{8}{13}, -\frac{1}{13}\right)$.

(b) The figure is below.

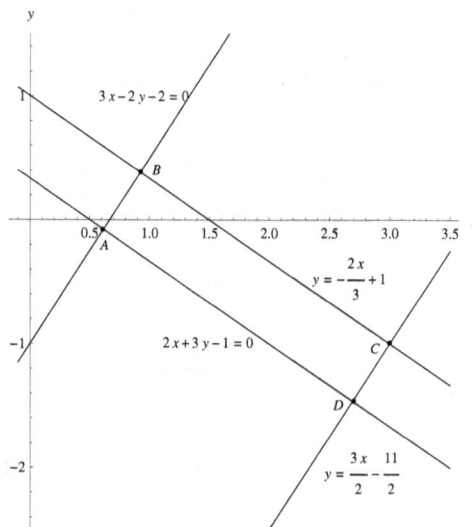

Figure 5.8: A plot of the lines for Exercise 11.

From Fig.(5.8), we see that the point $C(3, -1)$ must be on a straight line, say l_1, which is parallel to $2x + 3y - 1 = 0$ (slope $= -2/3$). Also, it must be on another straight line l_2 which is parallel to $3x - 2y - 2 = 0$ (slope $= 3/2$).

The equation of line l_1 can be written as $y = -\frac{2}{3}x + c_1$. Since point $C(3, -1)$ is on l_1, we can substitute point C to determine $c_1 = -1 + \frac{2}{3}(3) = 1$. So the equation of l_1 is $y = -\frac{2}{3}x + 1$. The intersection of l_1 and $3x - 2y - 2 = 0$ will give the coordinate of point B. Substitute $y = \frac{3x}{2} - 1$ into l_1:

$$l_1 : y = -\frac{2}{3}x + 1$$
$$\Rightarrow \frac{3x}{2} - 1 = -\frac{2}{3}x + 1$$
$$\frac{3x}{2} + \frac{2x}{3} = 2$$
$$\Rightarrow \frac{13x}{6} = 2$$
$$\therefore x = \frac{12}{13}$$
$$y = \frac{3(12/13)}{2} - 1$$
$$= \frac{18}{13} - 1 = \frac{5}{13}.$$

Thus point B is $\left(\frac{12}{13}, \frac{5}{13}\right)$.

Similarly, line l_2 can be written as $y = \frac{3}{2}x + c_2$. Using point C we obtain $c_2 = -1 - \frac{3}{2}(3) = -\frac{11}{2}$, and thus l_2 is $y = \frac{3}{2}x - \frac{11}{2}$. We can find the intersection of l_2 with $2x + 3y - 1 = 0$ to determine the point D. Substitute $y = -\frac{2}{3}x + \frac{1}{3}$ into l_2:

$$l_2 : y = \frac{3}{2}x - \frac{11}{2}$$
$$\Rightarrow -\frac{2}{3}x + \frac{1}{3} = \frac{3}{2}x - \frac{11}{2}$$
$$-\frac{2}{3}x - \frac{3}{2}x = -\frac{11}{2} - \frac{1}{3}$$
$$\Rightarrow -\frac{13x}{6} = -\frac{35}{6}$$
$$\therefore x = \frac{35}{13}$$
$$y = \frac{3(35/13)}{2} - \frac{11}{2}$$
$$= \frac{3(35) - 11(13)}{26} = -\frac{19}{13}.$$

Thus point D is $\left(\frac{35}{13}, -\frac{19}{13}\right)$.

(c) The lengths AB and AD are

$$l_{AB} = \sqrt{(x_A - x_B)^2 + (y_A - y_B)^2}$$
$$= \sqrt{\left(\frac{8}{13} - \frac{12}{13}\right)^2 + \left(-\frac{1}{13} - \frac{5}{13}\right)^2}$$
$$= \frac{2\sqrt{13}}{13}$$
$$l_{AD} = \sqrt{(x_A - x_D)^2 + (y_A - y_D)^2}$$
$$= \sqrt{\left(\frac{8}{13} - \frac{35}{13}\right)^2 + \left(-\frac{1}{13} + \frac{19}{13}\right)^2}$$
$$= \frac{9\sqrt{13}}{13}.$$

Thus the area of the rectangle is $A = l_{AB} \times l_{AD} = \frac{18}{13}$.

---

12. Given the points $A(1, a)$, $B(b, 2)$, $C(3, 5)$ and $D(4, 7)$.
(a) The straight line along CD is:

$$\frac{y - 5}{x - 3} = \frac{7 - 5}{4 - 3}$$
$$(y - 5) = 2x - 6$$
$$y = 2x - 1.$$

All points on this line, such as A and B, must satisfy the equation above. For A, $a = 2(1) - 1 = 1$, while for B, we have $2 = 2(b) - 1 \Rightarrow b = 3/2$.

(b) For AC to be perpendicular to CD, the product of their slope must be -1,

$$\frac{(y_C - y_A)}{(x_C - x_A)} \times \frac{(y_D - y_C)}{(x_D - x_C)} = -1$$
$$\Rightarrow \frac{5 - a}{3 - 1} \times \frac{7 - 5}{4 - 3} = -1$$
$$\frac{5 - a}{2} \times 2 = -1$$
$$5 - a = -1$$
$$a = 5 + 1 = 6.$$

Similarly, for AB to be perpendicular to AC,

$$\frac{(y_B - y_A)}{(x_B - x_A)} \times \frac{(y_C - y_A)}{(x_C - x_A)} = -1$$
$$\Rightarrow \frac{2 - 6}{b - 1} \times \frac{5 - 6}{3 - 1} = -1$$
$$-\frac{4}{b - 1} \times \frac{-1}{2} = -1$$
$$\frac{4}{b - 1} = -2$$
$$b - 1 = -2$$
$$b = -1.$$

42

(c) Given that A, B, C are collinear, their slopes must be the same,

$$\frac{y_B - y_A}{x_B - x_A} = \frac{y_C - y_B}{x_C - x_B} = \frac{y_C - y_A}{x_C - x_A}$$

$$\Rightarrow \quad \frac{2 - a}{b - 1} = \frac{3}{3 - b} = \frac{5 - a}{2}.$$

Also, we are given $b = a - 1$. After substitution we have

$$\frac{2 - a}{a - 2} = \frac{3}{4 - a} = \frac{5 - a}{2}.$$

Therefore $-1 = \frac{3}{4-a} \Rightarrow a - 4 = 3$. Thus $a = 7$; $b = a - 1 = 6$.

(d) Given that length AB is equal to length BC,

$$\sqrt{(x_B - x_A)^2 + (y_B - y_A)^2}$$
$$= \sqrt{(x_C - x_B)^2 + (y_C - y_B)^2}$$
$$\Rightarrow \quad (b - 1)^2 + (2 - a)^2 = (3 - b)^2 + (5 - 2)^2.$$

Then, given that $b = a + 1$, we get the following after substitution

$$a^2 + (2 - a)^2 = (2 - a)^2 + 3^2$$
$$a^2 = 9$$
$$\therefore a = \pm 3.$$

Thus we have $a = \pm 3$ with $b = 4, -2$ respectively.

(e) Given that the area of triangle ABD is 10.

$$10 = \frac{1}{2}\left| x_a(y_b - y_d) + x_b(y_d - y_a) \right.$$
$$\left. + x_d(y_a - y_b) \right|$$
$$10 = \frac{1}{2}\left| (2 - 7) + b(7 - a) + 4(a - 2) \right|$$
$$20 = \left| -5 + 7b - ab + 4a - 8 \right|$$
$$20 = \left| 7b + 4a - ab - 13 \right|.$$

As $a = 2b$,

$$20 = \left| -2b^2 + 15b - 13 \right|$$
$$20 = \begin{cases} (-2b^2 + 15b - 13) \\ -(-2b^2 + 15b - 13) \end{cases}$$

since $|x| = x$ for $x > 0$ and $|x| = -x$ for $x < 0$. Equivalently,

$$0 = \begin{cases} -2b^2 + 15b - 33 \\ 2b^2 - 15b - 7 \end{cases}$$

The first case does not give real solutions because the discriminant is $15^2 - 4(-2)(-33) =$

$-39 < 0$. The second case gives

$$b = \frac{15 \pm \sqrt{225 - 4(2)(-7)}}{2(2)}$$
$$= \frac{15 \pm \sqrt{281}}{4}.$$

Thus we have $a = 2b = (15 \pm \sqrt{281})/2$.

Did You Know?

Consider the sums:

$$S_1(n) = \sum_{k=1}^{n} k.$$

$$S_3(n) = \sum_{k=1}^{n} k^3.$$

Then

$$S_3(n) = [S_1(n)]^2.$$

For example,

$$1^3 + 2^3 = (1 + 2)^2.$$
$$1^3 + 2^3 + 3^3 = (1 + 2 + 3)^2.$$

Is that not wonderful?

5.2 Solutions to Odd-Numbered Problems

1. Plotting $\ln y$ against x^2, the straight line with slope m is $\ln y = mx^2 + c$. This line forms an angle of $30°$ with the horizontal axis, therefore $m = \tan 30° = \frac{\sqrt{3}}{3}$. Also, since $(x, y) = (0, 2)$ is a point on the line,

$$\frac{\ln y - \ln 2}{x^2 - 0} = \frac{\sqrt{3}}{3}$$

$$\ln \frac{y}{2} = \frac{x^2}{\sqrt{3}}$$

$$\Rightarrow y = 2e^{x^2/\sqrt{3}}.$$

---

3. (a) With two points $(0, 5)$ and $(2, 3)$ on the line, we can determine the equation of line as

$$\frac{y - 5}{x - 0} = \frac{5 - 3}{0 - 2}$$

$$\Rightarrow y = 5 - x.$$

Since $x = \sqrt{4 - z^2}$, by substitution we have $y = 5 - \sqrt{4 - z^2}$.

(b) We can rearrange the expression to obtain

$$y = 5 - \sqrt{4 - z^2}$$

$$\Rightarrow (5 - y)^2 = 4 - z^2$$

$$2^2 = (z - 0)^2 + (y - 5)^2.$$

This is the equation of a circle centred at $(0, 5)$ with radius $r = 2$. Also, since $y = 5 - \sqrt{4 - z^2} \leq 5$, the curve in part (a) is actually the lower half-circle, as shown below

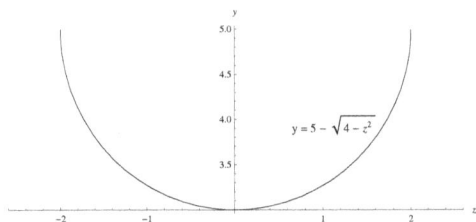

Figure 5.9: The sketch of semi-circle for Problem 3b.

---

5. (a) We need to solve the simultaneous equations by substitution. From $2x + 3y - 1 = 0 \Rightarrow y = (1 - 2x)/3$, so we have

$$2 = \left(\frac{1 - 2x}{3} + 1\right)(x + 3)$$

$$\Rightarrow 6 = (4 - 2x)(x + 3)$$

$$0 = x^2 + x - 3$$

$$\Rightarrow x = \frac{-1 \pm \sqrt{13}}{2}$$

Correspondingly, $y = \frac{2 \mp \sqrt{13}}{3}$. The points are $P\left(\frac{-1+\sqrt{13}}{2}, \frac{2-\sqrt{13}}{3}\right)$ and $Q\left(\frac{-1-\sqrt{13}}{2}, \frac{2+\sqrt{13}}{3}\right)$.

(b) The length between any two points on the graph is given by $D = \sqrt{(x_2 - x_1)^2 + (y_2 - y_1)^2}$. For P and Q,

$$D_{PQ} = \sqrt{(-\sqrt{13})^2 + (2\sqrt{13}/3)^2}$$

$$= \frac{13}{3}.$$

(c) The area of OPQ is

$$\text{Area} = \frac{1}{2}\big|x_O(y_P - y_Q) + x_P(y_Q - y_O)$$

$$+ x_Q(y_O - y_P)\big|$$

$$= \frac{1}{2}\left|\left(\frac{-1+\sqrt{13}}{2}\right)\left(\frac{2+\sqrt{13}}{3}\right)\right.$$

$$\left. - \left(\frac{-1-\sqrt{13}}{2}\right)\left(\frac{2-\sqrt{13}}{3}\right)\right|$$

$$= \frac{\sqrt{13}}{6}.$$

(d) The area of the triangle OPQ is given by $A = \frac{1}{2}D_{PQ}h$, where h is the shortest distance from O to PQ.

$$h = \frac{2A}{D_{PQ}} = \frac{2 \times \sqrt{13}/6}{13/3} = \frac{\sqrt{13}}{13}.$$

---

7. (a) We can rewrite the equation of the circle as

$$0 = x^2 + y^2 + 2x + 4y - 4$$

$$= [(x + 1)^2 - 1] + [(y + 2)^2 - 4] - 4$$

$$\Rightarrow 3^2 = (x + 1)^2 + (y + 2)^2.$$

This is a circle centred at $(-1, -2)$ with radius 3. Point $A(-1, 1)$ is on the circle, directly above the center $(-1, -2)$, while point $C(-1, -5)$ is directly below the center. Points B and D are on a line parallel to the x axis, with coordinates $(2,-2)$ and $(-4,-2)$.

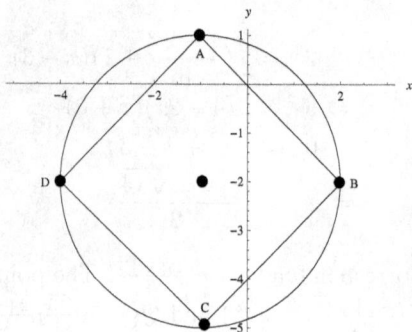

Figure 5.10: The sketch of circle and square $ABCD$ for problem 7.

(b) The length of AB is $\sqrt{(2-(-1))^2 + (-2-1)^2} = \sqrt{18}$. So the area of the square is 18.

---

9. (a) Let P be the point where the line $y = 5 - x$ is tangent to the circle of center $C(2,1)$. The line CP will be normal to the circle at point P. The slope of the normal is related to the slope of the tangent by $m_n = -1/m_t = 1$. Thus the equation of the normal is $y - 1 = 1(x - 2) \Rightarrow y = x - 1$. The intersection of the tangent and normal equations determines the coordinates of P. Solving the simultaneous equations, we get $P(3,2)$. Thus the radius is $R = D_{CP} = \sqrt{(3-2)^2 + (2-1)^2} = \sqrt{2}$.

(b) From part (a), $T_1 = P(3,2)$.

---

11. (a) The equation of the circle can be rewritten as $(x-1)^2 + (y-2)^2 = 3^2$. It is centred at $C(1,2)$. The slope of the tangent is $m = 2$, so we can write the equation of the tangent as $y = 2x + c$ where $c > 0$ is the positive y intercept. By substitution, the intercept satisfies

$$0 = 5x^2 + (4c - 10)x + c^2 - 4c - 4.$$

Since the tangent intercepts the circle at one point, the quadratic equation must have only one solution; its discriminant must be zero:

$$0 = (4c-10)^2 - 20(c^2 - 4c - 4)$$
$$\Rightarrow c = 3\sqrt{5}.$$

Thus, the equation of the tangent is $y = 2x + 3\sqrt{5}$.

(b) Substitute $y = 2x + 3\sqrt{5}$ into the equation of the circle,

$$0 = 5x^2 + (12\sqrt{5} - 10)x + 41 - 12\sqrt{5}$$
$$\Rightarrow x = \frac{5 - 6\sqrt{5}}{5}.$$

Then, $y = 2x + 3\sqrt{5} = (10 + 3\sqrt{5})/5$. Thus P is $\left(\frac{5-6\sqrt{5}}{5}, \frac{10+3\sqrt{5}}{5}\right)$.

(c) The slope of the normal is $m_n = -1/2$ and so its equation is

$$y - 2 = -\frac{1}{2}(x - 1)$$
$$\Rightarrow y = -\frac{1}{2}x + \frac{5}{2}.$$

(d) Given the points $O(0,0)$, $C(1,2)$ and $P\left(\frac{5-6\sqrt{5}}{5}, \frac{10+3\sqrt{5}}{5}\right)$, the area of triangle OPC can be determined using the formula in Problem (5c). The result is $3\sqrt{5}/2$.

(e) OP is $\sqrt{14}$. The equation of the circle is therefore $(x - x_p)^2 + (y - y_p)^2 = 14$ where (x_p, y_p) are the coordinates of P (see part (b)).

Did You Know?

Consider the infinite series

$$\sqrt{1 + \sqrt{1 + \sqrt{1 + \ldots}}}$$

It sums to $\dfrac{1 + \sqrt{5}}{2}$. Which is the Golden Ratio! (see Exercise (1) of Chap. 2).

Can you derive the result above?

(Hint: Let x equal the series and then square both sides to get $x^2 = 1 + x$.)

Chapter 6

Plane Geometry

6.1 Solutions to Exercises

1. (a) Proof of the Inscribed Angle Theorem. Consider the figure below, where D is a point on CO produced.

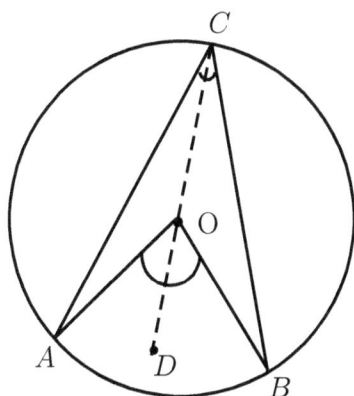

Figure 6.1: Schematic figure illustrating the Inscribed Angle Theorem. O is the centre of the circle.

Since $OA = OC$, $\triangle AOC$ is an isosceles triangle, and so $2\angle ACO + \angle AOC = \pi$. Also,

$$
\begin{aligned}
\angle AOD &= \pi - \angle AOC \\
&= \pi - (\pi - 2\angle ACO) = 2\angle ACO.
\end{aligned}
$$

Similarly, $\angle BOD = 2\angle BCO$. Therefore

$$
\begin{aligned}
\angle AOD + \angle BOD &= 2(\angle ACO) + \angle BCO) \\
\Rightarrow \beta &= 2\alpha \,,
\end{aligned}
$$

which is the Inscribed Angle Theorem.

Note: If C' is another point on the circumference on the same side of the chord AB as C, then this theorem also implies $\angle ACB = \angle AC'B$. That is, angles subtended by the same chord are equal, a useful fact in problem solving.

(b) Proof of the Tangent-Chord Theorem. Consider the figure below, where D is a point on the tangent line.

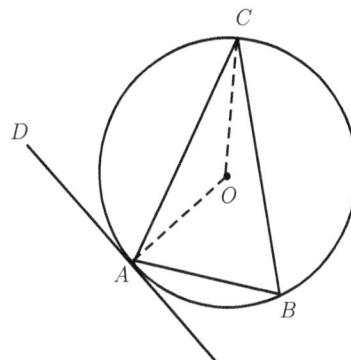

Figure 6.2: Schematic figure illustrating the Tangent-Chord Theorem. O is the centre of the circle.

By the Inscribed Angle Theorem (see part (a)), $\angle COA = 2\angle CBA$. Also, since OA and OC are the radii of the circle, $\triangle AOC$ is an isosceles triangle, which implies $\angle CAO = \frac{\pi - 2\alpha}{2} = \frac{\pi}{2} - \alpha$. Since OA is tangent to AD, we have $\angle DAC = \frac{\pi}{2} - \angle CAO = \alpha$. Hence we obtain the Tangent-Chord Theorem, $\angle DAC = \angle ABC$.

(c) Proof of the Intersecting Chord Theorem. Consider the figure below,

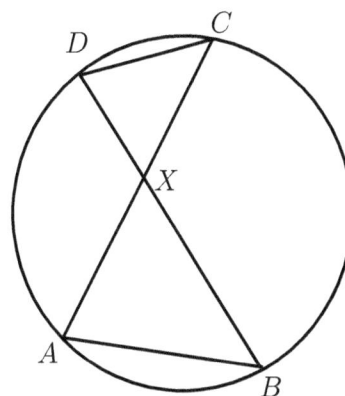

Figure 6.3: Schematic figure illustrating the Intersecting Chords Theorem.

By the Inscribed Angle Theorem, $\angle ABD = \angle ACD$ since they are angles subtended by the same chord AD. Similarly, $\angle BAC = \angle BDC$. Also, $\angle AXB = \angle CXD$ since they are pair of opposite angles. Thus we have the similar triangles $\triangle AXB \equiv \triangle CXD$. This is the Intersecting Chords Theorem, which can also be expressed as follows: From the similar triangles, we get the ratios $AB/CD = AX/DX = BX/CX$; therefore $(AX) \cdot (CX) = (BX) \cdot (DX)$.

(d) The Tangent-Secant Theorem. Consider the figure below,

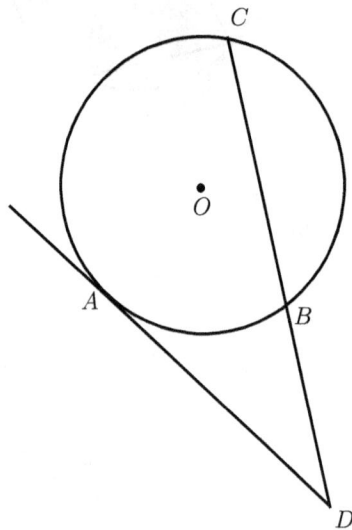

Figure 6.4: Figure illustrating the Tangent-Secant Theorem. AD is tangent to the circle.

As shown in Worked Example 4 of the book,

$$AD^2 = CD \times BD.$$

2. (a) Consider the figure below. Since the diameter AB passes through the origin O, by the Inscribed Angle Theorem we have

$$\alpha = 2\beta = \pi$$
$$\Rightarrow \beta = \frac{\pi}{2} = 90^0.$$

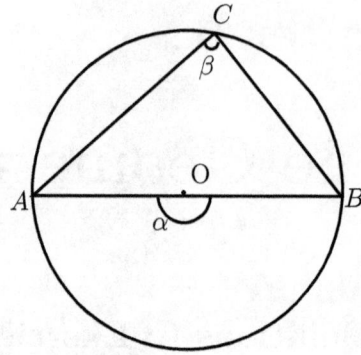

Figure 6.5: Schematic figure for Exercise 2(a).

(b) Consider triangle ABC in the figure below such that $\angle ACB = 90°$. AB is the hypotenuse of the triangle. By using the Inscribed Angle Theorem, angle AOB must be twice of angle C, that is $\alpha = 180°$.

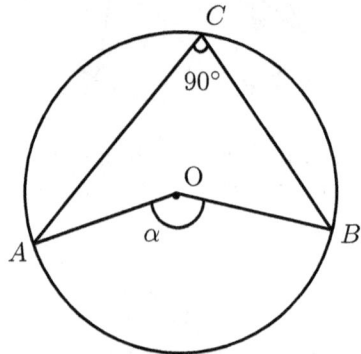

Figure 6.6: Schematic figure for Exercise 2(b).

Hence AOB is a straight line coinciding with the hypotenuse and the diameter of the circle.

3. Let us label the angles as shown in the figure below.

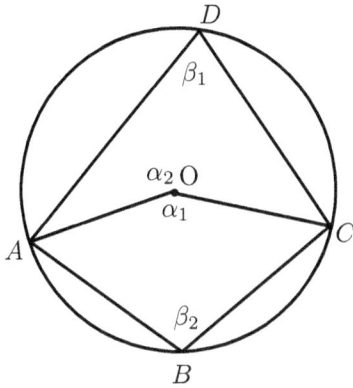

Figure 6.7: Exercise 3.

Consider the opposite vertices B and D. By using the Inscribed Angle Theorem, we have

$$\alpha_1 \;=\; 2\beta_1\,, \quad \alpha_2 = 2\beta_2.$$

Thus,

$$\beta_1 + \beta_2 \;=\; \frac{\alpha_1 + \alpha_2}{2}$$
$$=\; \frac{2\pi}{2} = \pi.$$

Since we did not use any unique properties of B and D, the result must be the same for the other pair of opposite vertices, C and A.

That is, the sum of the opposite interior angles of a cyclic quadrilateral is π.

4. Connect the points with lines, to form a triangle inscribed in a circle. Since $\angle ABC + \angle CAB + \angle BCA = \gamma + \beta + \alpha$ is the sum of interior angles in the triangle, it has to be π.

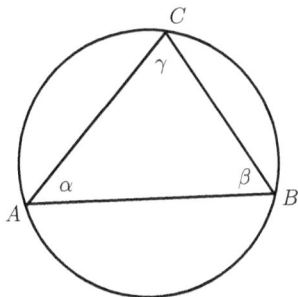

Figure 6.8: Exercise 4.

5. (a) Consider a right angled triangle inscribed in a circle. Since the hypotenuse is the longest

side, it must be the diameter of the circle, $D = 10$ (see also Q2b above).

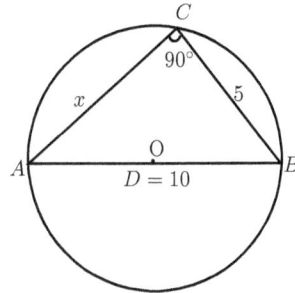

Figure 6.9: Exercise 5.

(b) By Pythagoras theorem, the length x of the other side of the triangle is determined by

$$(10)^2 \;=\; 5^2 + x^2$$
$$x \;=\; \sqrt{(10)^2 - 5^2} = 5\sqrt{3}.$$

6. (a) Since $OA=OD$, $\triangle AOD$ is isosceles. So $\angle CAD = \angle BDA = 25°$. Also, $\angle DAB$ is a right-angle since BD is the diameter. Thus, $\angle ABD = 90° - 25° = 65°$ as labelled in the figure.

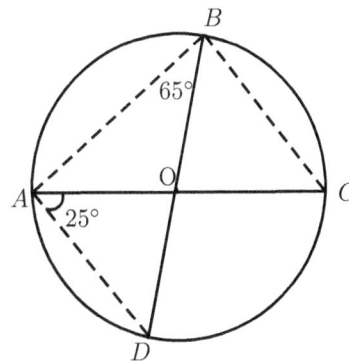

Figure 6.10: Schematic figure for Exercise 6.

(b) $\angle BCA = \angle BDA = 25°$ as they are angles subtended by the same chord (on the same side).

(c) $\angle COD = 2\angle CAD = 50°$, by the Inscribed Angle Theorem.

(d) Since BD is the diameter, $\angle BAD = 90°$ (see exercise 2(a) for proof).

7. Imagine a point P on the circumference so that EP is tangent to the circle.

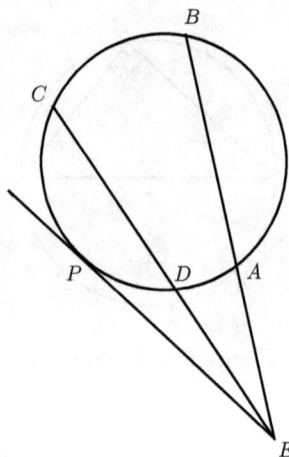

Figure 6.11: Exercise 7.

By the Tangent-Secant theorem, we have $EP^2 = EA \times EB = EC \times ED$. Thus, $EA = \frac{EC \times ED}{EB} = \frac{(8+5) \times 8}{11} = \frac{104}{11}$.

8. (a) Consider the figure below with $\theta = \angle ACB$,

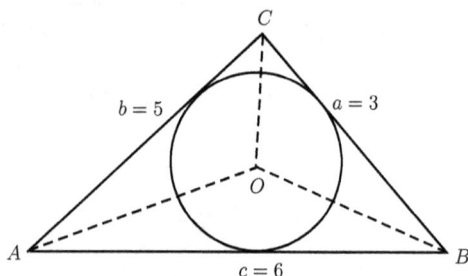

Figure 6.12: Exercise 8.

By using the cosine rule, we have

$$
\begin{aligned}
c^2 &= a^2 + b^2 - 2ab\cos\theta \\
\Rightarrow \cos\theta &= \frac{3^2 + 5^2 - 6^2}{2(3)(5)} = -\frac{1}{15}. \\
\therefore \sin\theta &= \frac{\sqrt{15^2 - 1}}{15} = \frac{4\sqrt{14}}{15}.
\end{aligned}
$$

The area of the triangle is thus $A = \frac{1}{2}ab\sin\theta = 2\sqrt{14}$.

(b) Since the radius r is perpendicular to each side, we can write down the areas of the following triangles: $A_{OCB} = \frac{1}{2}ra$, $A_{OAC} = \frac{1}{2}rb$ and $A_{OAB} = \frac{1}{2}rc$. The total area $A = \frac{1}{2}r(a + b + c) = 2\sqrt{14}$ (see part a). Thus, $r = \frac{4\sqrt{14}}{14} = \frac{2\sqrt{14}}{7}$.

9. (a) Consider the figure below,

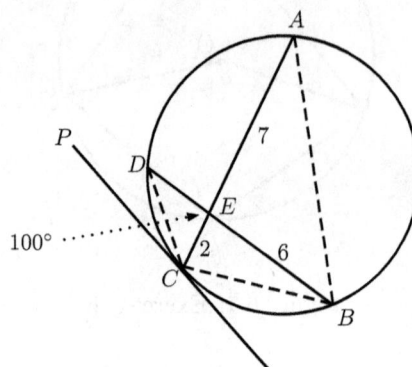

Figure 6.13: Exercise 9. $\angle DEC = 100°$

From the Intersecting-Chord Theorem, $AE \times EC = BE \times ED$. Therefore $ED = 7 \times 2/6 = 7/3$.

Then, applying the cosine rule to $\triangle CDE$,

$$
\begin{aligned}
DC^2 &= (ED)^2 + (EC)^2 \\
&\quad -2(EC)(ED)\cos 100° \\
DC &= \sqrt{\frac{7^2}{3^2} + 2^2 - 2\left(\frac{7}{3}\right)(2)\cos 100°} \\
&= \sqrt{11.065} = 3.3264 \approx 3.33.
\end{aligned}
$$

(b) $\angle ABD = \angle ECD$ by the Inscribed Angle Theorem. We determine $\angle ECD$ using the sine rule:

$$
\begin{aligned}
\frac{\sin \angle ECD}{ED} &= \frac{\sin 100°}{DC} \\
\angle ABD &= \angle ECD = \sin^{-1}\left(\frac{(7/3)\sin 100°}{3.3264}\right) \\
&= 43.69°.
\end{aligned}
$$

(c) Using the Tangent Chord Theorem,

$$
\begin{aligned}
\angle BCP &= \angle BAC = 180° - \angle ABD - \angle AEB \\
&= 180° - 43.69° - 100° = 36.31°.
\end{aligned}
$$

(Note: $\angle AEB = \angle DEC = 100°$.)

6.2 Solutions to Odd-Numbered Problems

1. Let the polygon have N vertices and draw lines from the centre of the circle to each vertex, dividing the polygon into N triangles as shown in Fig.(6.14). The sum of the interior vertex angles of the polygon is therefore $N\pi - 2\pi$ since each triangle contributes π less the amount at the centre.

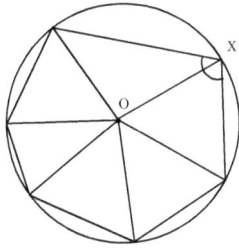

Figure 6.14: Problem 1. X is one of the interior vertex angles.

3. Consider the figure below,

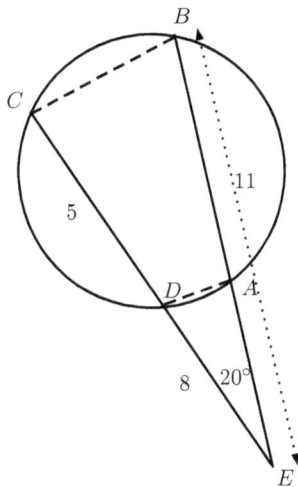

Figure 6.15: Problem 3.

BC follows from the cosine rule:

$$BC^2 = 13^2 + 11^2 - 2(13)(11)\cos(20°)$$
$$BC = 4.61.$$

We next determine EA using the Tangent Secant Theorem, which implies $EA \times 11 = ED \times EC$. Thus $EA = 104/11$.

Finally, AD is determined using the cosine rule:

$$AD^2 = 8^2 + (104/11)^2$$
$$-2(8)(104/11)\cos(20°)$$
$$AD = 3.35.$$

5. Consider the figure below,

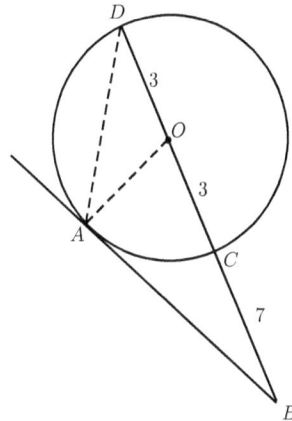

Figure 6.16: Problem 5.

From the right angle $\angle OAB$, simple trigonometry gives $\cos\angle AOB = 3/10$. From inscribed angle theorem, $\angle ADO = \angle AOB/2 = \frac{\cos^{-1}(3/10)}{2} = 36.27°$.

7. (a) Consider the figure below,

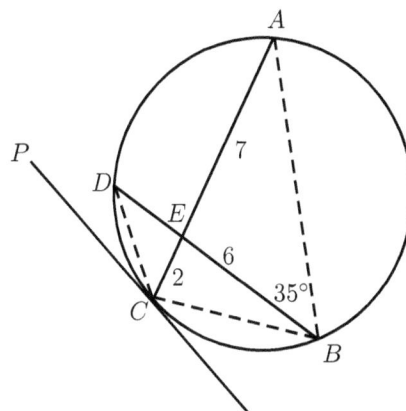

Figure 6.17: Problem 7.

$ED = 7/3$ as shown already in Exercise (9a).

We have $\angle ABD = \angle ACD = 35°$ as they are subtended by the same chord. Then, by using the cosine rule and solving the quadratic equation in DC, we have

$$
\begin{aligned}
ED^2 &= DC^2 + EC^2 - 2(DC)(EC)\cos 35° \\
\frac{49}{9} &= DC^2 + 4 - (4\cos 35°)DC \\
0 &= DC^2 - (4\cos 35°)DC - \frac{13}{9} \\
\Rightarrow DC &= 3.67 .
\end{aligned}
$$

(b) From $\triangle CDE$,

$$
\begin{aligned}
\frac{\sin\angle CED}{CD} &= \frac{\sin\angle ECD}{ED} \\
\angle CED &= \sin^{-1}\left(\frac{3.67\sin 35°}{7/3}\right) \\
&= 64.44° \ ; \ 115.56°.
\end{aligned}
$$

As $\angle CED$ is opposite the longest side, it must be the largest angle, so we have to choose $\angle CED = 115.56°$ and thus $\angle AED = 180° - 115.56° = 64.44°$. Next, using the cosine rule in $\triangle ADE$,

$$
\begin{aligned}
AD^2 &= AE^2 + ED^2 \\
&\quad -2(AE)(ED)\cos\angle AED \\
AD &= \sqrt{49 + \frac{49}{9} - \frac{98}{3}\cos 64.44°} \\
&= 6.3522.
\end{aligned}
$$

Finally, we use the sine rule in $\triangle ADE$,

$$
\begin{aligned}
\frac{\sin\angle ADE}{7} &= \frac{\sin 64.44°}{6.3522} \\
\Rightarrow \angle ADE &= 83.79° \ ; \ 96.21°.
\end{aligned}
$$

As before, we have to choose $\angle ADE \approx 96.2°$.

(c) From $\triangle ADE$, $\angle CAD = 180° - \angle AED - \angle ADE = 180° - 64.44° - 96.21° = 19.35°$. By tangent-chord theorem, $\angle DCQ = \angle CAD = 19.35°$. Thus, the angle between the tangent at C and the line EC is given by $\angle ACP = 180° - \angle ACD - \angle DCQ = 180° - 35° - 19.35° \approx 126°$.

9. Consider the figure below,

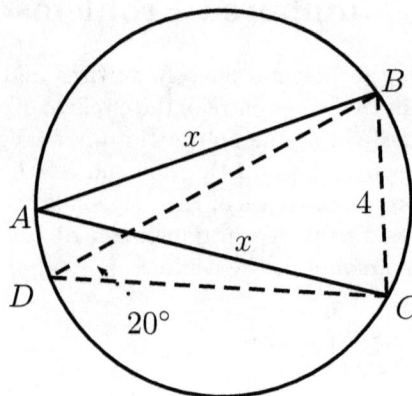

Figure 6.18: Problem 9.

$\angle BAC = \angle BDC = 20°$ as they are subtended by the same chord. Using the cosine rule in $\triangle BAC$,

$$
\begin{aligned}
BC^2 &= AB^2 + AC^2 \\
&\quad -2(AB)(AC)\cos\angle BAC \\
4^2 &= 2x^2(1 - \cos 20°) \\
\Rightarrow x &= \frac{4}{\sqrt{2(1 - \cos 20°)}} .
\end{aligned}
$$

Thus the area of triangle ABC is

$$
\begin{aligned}
\text{Area} &= \frac{1}{2}x^2\sin\angle BAC \\
&= \frac{1}{2}\times\left(\frac{16}{2(1-\cos 20°)}\right)\sin 20° \\
&= 22.69.
\end{aligned}
$$

Did You Know?

The number 10^{100} is called a "googol". Is it related to the web search engine Google? Well, 'google' to find out!

Chapter 7

Differential Calculus

7.1 Solutions to Exercises

1. *Note: Velocity is the rate of change of displacement with time. Its components are $v_x = \frac{dx}{dt}$, and so on.*

   From $E(3)$ of chapter 4, we have $x(t) = Ut\cos\theta$, $y(t) = Ut\sin\theta - gt^2/2$. So the velocity components of the projectile in the x and y directions are,

   $$v_x = \frac{dx}{dt} = U\cos\theta \ ,$$

   $$v_y = \frac{dy}{dt} = U\sin\theta - gt.$$

2. Given that $f(x) = x^n$, $g(x) = x^m$.
   (a) To verify the Product Rule,

   $$\frac{d}{dx}(fg) = \frac{d}{dx}(x^n \cdot x^m) = \frac{d}{dx}(x^{m+n})$$
   $$= (m+n)x^{m+n-1}.$$

   Meanwhile,

   $$f\frac{dg}{dx} + g\frac{df}{dx} = x^n\frac{dx^m}{dx} + x^m\frac{dx^n}{dx}$$
   $$= x^n(mx^{m-1}) + x^m(nx^{n-1})$$
   $$= mx^{n+m-1} + nx^{m+n-1}$$
   $$= (m+n)x^{m+n-1}.$$

   Thus we verified the product rule $\frac{d}{dx}(fg) = f\frac{dg}{dx} + g\frac{df}{dx}$.

   (b) To verify the Quotient Rule,

   $$\frac{d}{dx}\frac{f}{g} = \frac{d}{dx}\frac{x^n}{x^m} = \frac{d}{dx}(x^{n-m})$$
   $$= (n-m)\,x^{n-m-1}.$$

   Meanwhile,

   $$\frac{gf' - fg'}{g^2} = \frac{x^m \cdot (nx^{n-1}) - x^n \cdot (mx^{m-1})}{x^{2m}}$$
   $$= \frac{(n-m)\,x^{m+n-1}}{x^{2m}}$$
   $$= (n-m)\,x^{n-m-1}.$$

Thus the Quotient rule is verified.

(c) To verify the Chain Rule,

$$\frac{d}{dx}f(g(x)) = \frac{d}{dx}f(x^m) = \frac{d}{dx}(x^m)^n$$
$$= \frac{d}{dx}x^{mn} = mn\ x^{mn-1}.$$

Meanwhile,

$$\frac{df}{dg} \times \frac{dg}{dx} = \frac{d}{dg}(g^n) \times \frac{d}{dx}x^m$$
$$= ng^{n-1} \times mx^{m-1} = mn(x^m)^{n-1} \times x^{m-1}$$
$$= mn\ x^{mn-m} \times x^{m-1} = mn\ x^{mn-1}.$$

Thus the Chain Rule is verified.

(d)

$$\frac{df}{dx} = nx^{n-1}\ .$$
$$\left(\frac{dx}{df}\right)^{-1} = \left(\frac{d}{df}(f^{1/n})\right)^{-1}$$
$$= \left(\frac{1}{n}f^{(\frac{1}{n}-1)}\right)^{-1}$$
$$= n\ f^{(1-\frac{1}{n})} = n\ (x^n)^{(1-\frac{1}{n})}$$
$$= n\ x^{n-1}.$$

Hence $\frac{df}{dx} = \left(\frac{dx}{df}\right)^{-1}$ is verified.

3. (a) *Recall $\frac{d}{dx}(Ax^n) = Anx^{n-1}$.*

   $$\frac{d}{dx}[3x^2 + 2x^{-3}] = 3(2)x^{2-1} + 2(-3)x^{-3-1}$$
   $$= 6x - 6x^{-4}.$$

   (b) *Recall $\frac{d}{dx}\ln(f(x)) = \frac{f'(x)}{f(x)}$.*

   $$\frac{d}{dx}[\ln(2x+1)] = \frac{1}{2x+1} \times \frac{d}{dx}(2x+1)$$
   $$= \frac{2}{2x+1}.$$

52

(c) Recall $\frac{d}{dx}\sin(f(x)) = \cos(f(x)) \times f'(x)$.

$$\frac{d}{dx}[2\sin(3x)] = 2\cos(3x) \times \frac{d}{dx}(3x)$$
$$= 2(3)\cos(3x) = 6\cos(3x).$$

(d) Recall $\frac{d}{dx}\cos(f(x)) = -\sin(f(x)) \times f'(x)$.
We let $u = x^2$; $v = \cos x \Rightarrow \cos^2 x = u(v(x))$.

$$\frac{d}{dx}[3\cos^2 x] = 3\frac{d}{dx}u(v(x))$$
$$= 3\frac{du}{dv} \times \frac{dv}{dx} = 3(2v) \times (-\sin x)$$
$$= -6\cos x \sin x = -3\sin(2x).$$

(e) Recall the product rule, $\frac{d}{dx}(fg) = f\frac{dg}{dx} + g\frac{df}{dx}$.
Let $f(x) = x$; $g(x) = e^{2x}$,

$$\frac{d}{dx}[3xe^{2x}] = 3(f\frac{dg}{dx} + g\frac{df}{dx})$$
$$= 3(x(2e^{2x}) + e^{2x}(1))$$
$$= 3e^{2x}(2x+1).$$

(f) Recall the quotient rule, $\frac{d}{dx}\frac{f}{g} = \frac{gf' - fg'}{g^2}$.
Let $f(x) = 2 + 3x$; $g(x) = x + 5$,

$$\frac{d}{dx}\left[\frac{2+3x}{x+5}\right] = \frac{gf' - fg'}{g^2}$$
$$= \frac{(x+5)(3) - (2+3x)}{(x+5)^2}$$
$$= \frac{3x + 15 - 2 - 3x}{(x+5)^2}$$
$$= \frac{13}{(x+5)^2}.$$

---

4. Given that $y = f(x)$ and $x = x(t)$ such that $\frac{dx}{dt} = 2$ at $x = 0$. By using the Chain Rule

$$\frac{dy}{dt} = \frac{dy}{dx} \times \frac{dx}{dt}$$
$$= 2 \times \frac{dy}{dx}\Big|_{x=0}.$$

By using the results of Question 3, we have:

(a)

$$\frac{d}{dt}[3x^2 + 2x^{-3}]$$
$$= 2 \times (6x - 6x^{-4})\Big|_{x\to 0} \longrightarrow -\infty.$$

(b)

$$\frac{d}{dt}[\ln(2x+1)] = 2 \times \left(\frac{2}{2x+1}\right)\Big|_{x=0}$$
$$= 2 \times 2 = 4.$$

(c)

$$\frac{d}{dt}[2\sin(3x)] = 2 \times (6\cos(3x))\Big|_{x=0}$$
$$= 2 \times 6 = 12.$$

(d)

$$\frac{d}{dt}[3\cos^2 x] = 2 \times (-3\sin(2x))\Big|_{x=0}$$
$$= 2 \times 0 = 0.$$

(e)

$$\frac{d}{dt}[3xe^{2x}] = 2 \times (3e^{2x}(2x+1))\Big|_{x=0}$$
$$= 2 \times 3 = 6.$$

(f)

$$\frac{d}{dt}\left[\frac{2+3x}{x+5}\right] = 2 \times \left(\frac{13}{(x+5)^2}\right)\Big|_{x=0}$$
$$= 2 \times \frac{13}{25} = \frac{26}{25}.$$

---

5. Using the identities and rules given in Section 7.2 of the book, we can evaluate the following:
(a) Apply $\frac{d}{dx}(Ax^n) = nAx^{n-1}$,

$$\frac{d}{dx}\left[3x^2(1+x) + \frac{2}{x}\right]$$
$$= 3x^2\frac{d}{dx}(1+x) + (1+x)\frac{d}{dx}(3x^2) + \frac{d}{dx}\left(\frac{2}{x}\right)$$
$$= 3x^2 + (1+x)6x + 2(-1)x^{-2}$$
$$= 3x^2 + 6x^2 + 6x - 2x^{-2}$$
$$= 9x^2 + 6x - 2x^{-2}.$$

(b) Apply the Product Rule,

$$\frac{d}{dx}\left[\sqrt{2+3x}\right]$$
$$= \frac{1}{2}(2+3x)^{\frac{1}{2}-1} \times \frac{d}{dx}(2+3x)$$
$$= \frac{1}{2}(2+3x)^{-1/2} \times 3$$
$$= \frac{3}{2\sqrt{2+3x}}.$$

(c) Apply the Quotient Rule,

$$\frac{d}{dx}\left[\frac{\sqrt{2+3x}}{x^3+5}\right]$$

$$= \frac{(x^3+5)\frac{d}{dx}(\sqrt{2+3x}) - \sqrt{2+3x}\frac{d}{dx}(x^3+5)}{(x^3+5)^2}$$

$$= \frac{\frac{3(x^3+5)}{2\sqrt{2+3x}} - 3x^2\sqrt{2+3x}}{(x^3+5)^2}$$

$$= \frac{3(x^3+5) - 6x^2(2+3x)}{2\sqrt{2+3x}(x^3+5)^2}$$

$$= \frac{3x^3+15 - 12x^2 - 18x^3}{2\sqrt{2+3x}(x^3+5)^2}$$

$$= \frac{-15x^3 - 12x^2 + 15}{2\sqrt{2+3x}(x^3+5)^2}$$

$$= \frac{-3(5x^3+4x^2-5)}{2\sqrt{2+3x}(x^3+5)^2}.$$

(d)

$$\frac{d}{dx}\left[\sin 2x + x^2\cos x + x\tan 5x\right]$$

$$= \frac{d}{dx}\sin 2x + x^2\frac{d}{dx}\cos x + \cos x\frac{d}{dx}x^2$$

$$+ x\frac{d}{dx}\tan 5x + \tan 5x\frac{d}{dx}x$$

$$= 2\cos 2x - x^2\sin x + 2x\cos x$$

$$+ 5x\sec^2 5x + \tan 5x.$$

(e)

$$\frac{d}{dx}\left[x^2 e^x + \ln(1+x)\right]$$

$$= x^2\frac{d}{dx}e^x + e^x\frac{d}{dx}x^2 + \frac{1}{(1+x)}\frac{d}{dx}(1+x)$$

$$= x^2 e^x + 2xe^x + \frac{1}{(1+x)}$$

$$= x(x+2)e^x + \frac{1}{1+x}.$$

(f)

$$\frac{d}{dx}\left[\sin(\ln x)\right] = \cos(\ln x)\frac{d}{dx}\ln x$$

$$= \frac{\cos(\ln x)}{x}.$$

6. (a) Since both a and p are positive constants, the first term on the right hand side is positive, $aV^p > 0$. It will lead to an increase of volume $V(t)$, and so it is the growth term. Similarly, the second term, $-bV^q < 0$, is the degradation term.

(b) The tumour will grow in size when,

$$\frac{dV}{dt} = aV^p - bV^q > 0$$

$$\Rightarrow aV^p > bV^q$$

$$\frac{V^p}{V^q} = V^{p-q} > \frac{b}{a}$$

$$\therefore V > \left(\frac{b}{a}\right)^{1/(p-q)}.$$

---

7. (a) Given $y(x) = ax^2 + bx + c$, the derivatives are:

$$\frac{dy}{dx} = \frac{d}{dx}\left[ax^2 + bx + c\right]$$

$$= 2ax + b$$

$$\frac{d^2y}{dx^2} = \frac{d}{dx}\left[2ax + b\right]$$

$$= 2a.$$

(b) For stationary points, we let $\frac{dy}{dx} = 0$, so

$$\frac{dy}{dx} = 0 \Rightarrow 2ax + b = 0$$

$$\therefore x = -\frac{b}{2a}.$$

Since $a \neq 0$, there is always a stationary point situated at $x = -\frac{b}{2a}$. The nature of the turning point can be determined by looking at the second derivative, $\frac{d^2y}{dx^2} = 2a$. For $a > 0 \Rightarrow \frac{d^2y}{dx^2} > 0$, we have a local minimum. Similarly, for $a < 0 \Rightarrow \frac{d^2y}{dx^2} < 0$ the turning point is a local maximum.

A plot of the quadratic function $y(x) = ax^2 + bx + c$ shows that when $a > 0$ the turning point is in fact a global minimum. Similarly, for $a < 0$ we have a global maximum.

(c) Completing the square,

$$y = ax^2 + bx + c = a\left[\frac{x^2}{a} + \frac{b}{a}x + \frac{c}{a}\right]$$

$$= a\left[\left(x + \frac{b}{2a}\right)^2 - \frac{b^2}{4a^2} + \frac{c}{a}\right]$$

$$= a\left(x + \frac{b}{2a}\right)^2 + c - \frac{b^2}{4a}.$$

If $a > 0$, then the first term (in the last equation) is non-negative, achieving a minimum at $x = -b/2a$, with $y_{\min} = c - \frac{b^2}{4a}$.

Similarly, if $a < 0$, there is a maximum at $x = -\frac{b}{2a}$ with $y_{\max} = c + \frac{b^2}{4|a|}$.

8. (a) Given that $y(x) = 2x + \frac{1}{1+x}$. The derivatives are:

$$\frac{dy}{dx} = \frac{d}{dx}\left[2x + \frac{1}{1+x}\right]$$

$$= 2 + (-1)(1+x)^{-2} = 2 - \frac{1}{(1+x)^2};$$

$$\frac{d^2y}{dx^2} = \frac{d}{dx}\left[2 - \frac{1}{(1+x)^2}\right]$$

$$= -(-2)(1+x)^{-3} = \frac{2}{(1+x)^3}.$$

For stationary points, we set

$$\frac{dy}{dx} = 2 - \frac{1}{(1+x)^2} = 0$$

$$\Rightarrow \frac{1}{(1+x)^2} = 2$$

$$(1+x)^2 = \frac{1}{2}$$

$$1+x = \pm\frac{1}{\sqrt{2}}$$

$$\therefore x = \pm\frac{1}{\sqrt{2}} - 1.$$

Let $x_1 = \frac{1}{\sqrt{2}} - 1$ and $x_2 = -\frac{1}{\sqrt{2}} - 1$. Since $\frac{d^2y}{dx^2} = \frac{2}{(1+x)^3}$ is positive for $x > -1$ and negative for $x < -1$, we have

$$\left.\frac{d^2y}{dx^2}\right|_{x=x_1} > 0 \Rightarrow x_1 \text{ (local minimum)};$$

$$\left.\frac{d^2y}{dx^2}\right|_{x=x_2} < 0 \Rightarrow x_2 \text{ (local maximum)}.$$

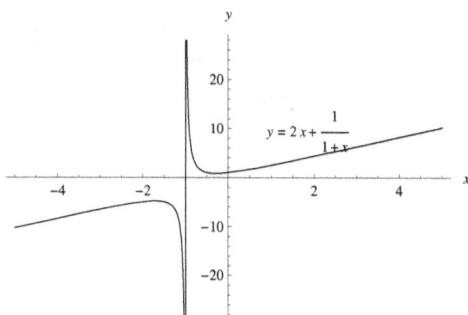

Figure 7.1: Plot of $y(x) = 2x + \frac{1}{1+x}$.

From the graph, we see that the extrema are respectively a local minimum and a local maximum. (The point $x = -1$ is a singular point.)

(b) Given $y(x) = \sqrt{2 + 3x^2}$, the derivatives are:

$$\frac{dy}{dx} = \frac{d}{dx}\left[\sqrt{2 + 3x^2}\right]$$

$$= \frac{1}{2}(2 + 3x^2)^{-1/2}(6x) = \frac{3x}{\sqrt{2+3x^2}};$$

$$\frac{d^2y}{dx^2} = \frac{d}{dx}\left[\frac{3x}{\sqrt{2+3x^2}}\right]$$

$$= \frac{\sqrt{2+3x^2}(3) - \frac{3x}{2}(2+3x^2)^{-1/2}(6x)}{(\sqrt{2+3x^2})^2}$$

$$= \frac{3\sqrt{2+3x^2} - 9x^2(2+3x^2)^{-1/2}}{2+3x^2}$$

$$= \frac{2}{\sqrt{2+3x^2}} - \frac{9x^2}{(2+3x^2)^{3/2}}.$$

For stationary points, we set $\frac{dy}{dx} = \frac{3x}{\sqrt{2+3x^2}} = 0 \Rightarrow x = 0$. At the extremum, the second derivative is

$$\left.\frac{d^2y}{dx^2}\right|_{x=0} = \left[\frac{2}{\sqrt{2+3x^2}} - \frac{9x^2}{(2+3x^2)^{3/2}}\right]\Big|_{x=0}$$

$$= \frac{3}{\sqrt{2}} > 0$$

$$\Rightarrow \quad \text{local minimum at } x = 0.$$

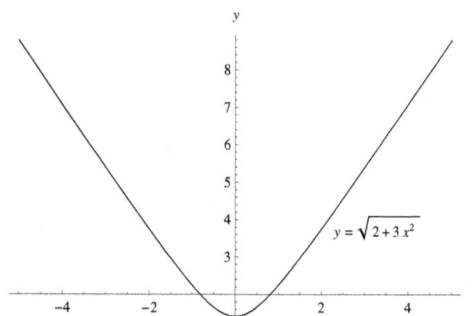

Figure 7.2: Plot of $y(x) = \sqrt{2 + 3x^2}$.

From the graph, we see that the extremum at $x = 0$ is actually a global minimum rather than just a local minimum.

(c) Given $y(x) = (x+1)^2 e^{-x}$, the derivatives

are:

$$\begin{aligned}\frac{dy}{dx} &= \frac{d}{dx}\left[(x+1)^2 e^{-x}\right]\\ &= 2(x+1)e^{-x} + (x+1)^2(-e^{-x})\\ &= [2x+2-(x^2+2x+1)]e^{-x}\\ &= (1-x^2)e^{-x};\\ \frac{d^2y}{dx^2} &= \frac{d}{dx}\left[(1-x^2)e^{-x}\right]\\ &= -2xe^{-x} + (1-x^2)(-e^{-x})\\ &= (x^2-2x-1)e^{-x}.\end{aligned}$$

For stationary points, we set $\frac{dy}{dx} = (1-x^2)e^{-x} = 0 \Rightarrow x = \pm 1$. At the extrema, the second derivatives are

$$\left.\frac{d^2y}{dx^2}\right|_{x=1} = -2e^{-1} < 0$$
$$\Rightarrow \text{ a local maximum at } x = 1.$$
$$\left.\frac{d^2y}{dx^2}\right|_{x=-1} = 2e > 0$$
$$\Rightarrow \quad \text{a local minimum at } x = -1.$$

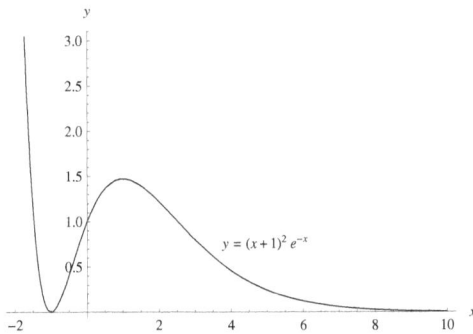

Figure 7.3: Plot of $y(x) = (x+1)^2 e^{-x}$.

From the graph, we see that the extremum at $x = -1$ is actually a global minimum while the extremum at $x = 1$ is indeed a local maximum.

(d) Given $y(x) = \sin x^2 + x^2$ for $x^2 \leq \pi$, the derivatives are:

$$\begin{aligned}\frac{dy}{dx} &= \frac{d}{dx}\left[\sin x^2 + x^2\right]\\ &= 2x\cos x^2 + 2x\\ &= 2x(\cos x^2 + 1);\\ \frac{d^2y}{dx^2} &= \frac{d}{dx}\left[2x(\cos x^2 + 1)\right]\\ &= 2(\cos x^2 + 1) + 2x(-\sin x^2)(2x)\\ &= 2\cos x^2 - 4x^2\sin x^2 + 2.\end{aligned}$$

For stationary points, we set $\frac{dy}{dx} = 2x(\cos x^2 + 1) = 0 \Rightarrow x = 0$, or $\cos x^2 = -1$. Since $x^2 \leq$

π, we have $x = 0, \sqrt{\pi}$. At stationary points, the second derivatives are

$$\left.\frac{d^2y}{dx^2}\right|_{x=0} = 4 > 0$$
$$\Rightarrow \text{ local minimum at } x = 0.$$
$$\left.\frac{d^2y}{dx^2}\right|_{x=\sqrt{\pi}} = 0 \Rightarrow \text{inconclusive.}$$

The second derivative test is inconclusive for $x = \sqrt{\pi}$. To determine the nature of this stationary point, we have to examine the first derivative in its vicinity.

x	$x=\left(\sqrt{\pi}\right)^-$	$x=\sqrt{\pi}$	$x=\left(\sqrt{\pi}\right)^+$
$\frac{dy}{dx}$	+	0	+

Figure 7.4: First derivative test at $x = \sqrt{\pi}$.

From the table, we see that the slope dy/dx before and after the point $x = \sqrt{\pi}$ has the same sign. Hence we conclude that $x = \sqrt{\pi}$ is an inflexion point.

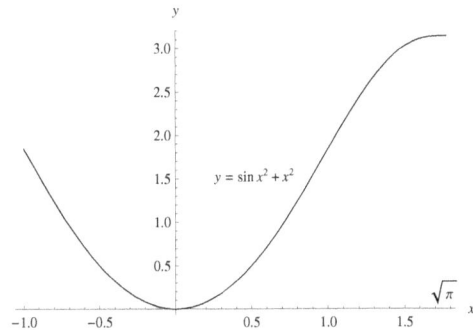

Figure 7.5: Plot of $y(x) = \sin x^2 + x^2$ for $x \leq \sqrt{\pi}$.

From the graph, we see that the stationary point at $x = 0$ is a global minimum while the point $x = \sqrt{\pi}$ is an inflexion point.

(e) Given $y(x) = \ln(x^2 + x + 3)$, the derivatives

are:

$$\frac{dy}{dx} = \frac{d}{dx}\left[\ln(x^2 + x + 3)\right]$$

$$= \frac{1}{x^2 + x + 3} \times \frac{d}{dx}(x^2 + x + 3)$$

$$= \frac{2x + 1}{x^2 + x + 3};$$

$$\frac{d^2y}{dx^2} = \frac{d}{dx}\left[\frac{2x + 1}{x^2 + x + 3}\right]$$

$$= \frac{2(x^2 + x + 3) - (2x + 1)(2x + 1)}{(x^2 + x + 3)^2}$$

$$= \frac{-2x^2 - 2x + 5}{(x^2 + x + 3)^2}.$$

For the stationary points, we set $\frac{dy}{dx} = \frac{2x+1}{x^2+x+3} = 0 \Rightarrow x = -\frac{1}{2}$. At the extremum, the second derivative is

$$\frac{d^2y}{dx^2}\bigg|_{x=-1/2} = \frac{8}{11} > 0$$

$$\Rightarrow x = -1/2 \text{ (local minimum)}.$$

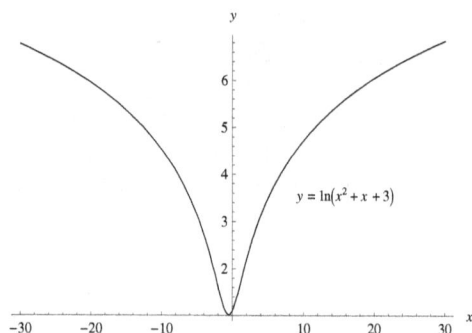

Figure 7.6: Plot of $y(x) = \ln(x^2 + x + 3)$.

From the graph, we see that the extremum at $x = -1/2$ is a global minimum.

9. *Note: If the tangent is $y = mx + c_1$, then the normal is $y = -\frac{1}{m}x + c_2$. Here m is the slope of the tangent and c_1, c_2 are constants.*

(a) Given that $y = x^2 - 3x + 5$, the slope at the point $x = 1$ (and $y = 3$) on the curve is

$$\frac{dy}{dx} = 2x - 3\big|_{x=1} = 2(1) - 3 = -1.$$

At $x = 1$, the tangent equation is $y = -x + c_1$. Since the point $(1, 3)$ must be on the tangent

and normal lines at $x = 1$, we can use it to determine the constants c_1, c_2. Substitution gives $c_1 = 3 + 1 = 4$, and thus the equation of the tangent is $y = -x + 4$. Similarly, for the normal $y = -\frac{1}{m}x + c_2 = x + c_2$, we can fix $c_2 = 3 - 1 = 2$. Thus the equation of normal is $y = x + 2$.

(b) Given that $y = x^3 + x^2 + 2x + 1$, the slope at the point $x = 1$ (and $y = 5$) on the curve is

$$\frac{dy}{dx} = 3x^2 + 2x + 2\big|_{x=1}$$

$$= 3(1) + 2(1) + 2 = 7.$$

At $x = 1$, the tangent equation is $y = 7x + c_1$. As the point $(1, 5)$ is on the tangent and normal lines at $x = 1$, we can use it to determine constants c_1, c_2. Substitution gives $c_1 = 5 - 7(1) = -2$, and thus equation of tangent, $y = 7x - 2$. Similarly, for the normal $y = -\frac{1}{m}x + c_2 = -\frac{1}{7}x + c_2$, we can fix $c_2 = 5 + \frac{1}{7} = \frac{36}{7}$. Thus the equation of normal is $y = \frac{-x+36}{7}$.

10. (a) *Note: To examine whether a function is increasing or decreasing with respect to x, we need to check the slope of the function. For instance, if $dy/dx > 0$ for $x > a$, then the function increases in the domain $x > a$.*

(i) The slope in Question 9(a) is $y' = 2x - 3$. Therefore the function decreases with increasing x when $y' < 0 \Rightarrow$ when $x < \frac{3}{2}$.

(ii) The slope in Question 9(b) is $y' = 3x^2 + 2x + 2$ and similar to part (i) we require $y' < 0$. Completing the square, we obtain

$$y' = 3\left[x^2 + \frac{2x}{3} + \frac{2}{3}\right]$$

$$= 3\left[\left(x + \frac{1}{3}\right)^2 - \frac{1}{3^2} + \frac{2}{3}\right]$$

$$= 3\left(x + \frac{1}{3}\right)^2 + \frac{5}{3} > 0.$$

Hence $y' < 0$ is impossible. There is no region where y decreases as x increases.

(b) *Note: To examine whether the slope is increasing or decreasing with respect to x, we need to check the second derivative of the function. For instance, if $d^2y/dx^2 > 0$ for $x > a$, then the slope of the function increases* *for $x > a$.*

(i) The second derivative for Question 9(a) is $y'' = 2$. To check for increasing y', we require $y'' > 0$. Since $y'' = 2$ for $x \in R$, the slope increases for all real $x \in (-\infty, \infty)$.

(ii) The second derivative for Question 9(b) is $y'' = 6x + 2$.
$y'' > 0 \Rightarrow 6x + 2 > 0$, or $x > -\frac{1}{3}$. Hence, the slope increases for all real $x \in \left(-\frac{1}{3}, \infty\right)$.

_____

11. (a) For stationary points,

$$\frac{dx}{dt} = 0 \Rightarrow Ax(B - x) = 0.$$

Hence $x = 0$ or B.

(b) When $x > B$, $\frac{dx}{dt} < 0$. It implies a declining population.

(c)

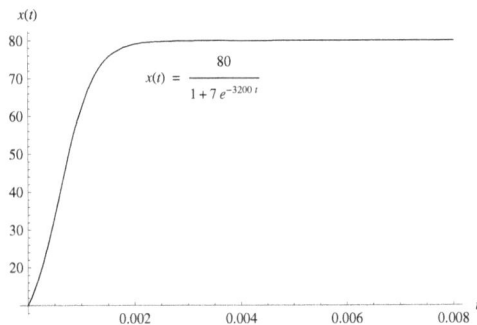

Figure 7.7: The logistic curve rises exponentially at first but then slows down, reaching an asymptotic value.

(d) B is the asymptotic value of the population, reached when $t \to \infty$. It is also a stable equilibrium point.

_____

12. Using the Chain Rule, we can determine the following:
(a) Circumference of the great circle, $C(R) = 2\pi R$ and its rate of change is

$$\frac{dC(R)}{dt} = \frac{dC(R)}{dR} \times \frac{dR}{dt}$$
$$= k\frac{d}{dR}(2\pi R) = 2\pi k \text{ cm/s}.$$

(b) Surface area of the sphere, $A(R) = 4\pi R^2$ and its rate of change is

$$\frac{dA(R)}{dt} = \frac{dA(R)}{dR} \times \frac{dR}{dt}$$
$$= k\frac{d}{dR}(4\pi R^2) = 8\pi kR \text{ cm}^2/\text{s}.$$

(c) Volume of the sphere, $V(R) = \frac{4}{3}\pi R^3$ and its rate of change is

$$\frac{dV(R)}{dt} = \frac{dV(R)}{dR} \times \frac{dR}{dt}$$
$$= k\frac{d}{dR}(\frac{4}{3}\pi R^3) = 4\pi kR^2 \text{ cm}^3/\text{s}.$$

_____

13. *Note: The approximate change of a function is $\Delta y \approx \left(\frac{dy}{dx}\right) \Delta x$. Given that the radius of sphere changes from 10 cm to 10.1 cm, so $\Delta R = 0.1$ cm.*

(a) Since $C(R) = 2\pi R$ and $\frac{dC(R)}{dR} = 2\pi$,

$$\Delta C \approx \frac{dC(R)}{dR} \times \Delta R = 2\pi \Delta R$$
$$= 2\pi \times 0.1 \text{ cm}$$
$$= 0.2\pi \text{ cm} = 0.628 \text{ cm}.$$

(b) The surface area is $A(R) = 4\pi R^2$ and $\frac{dA(R)}{dR} = 8\pi R$,

$$\Delta A \approx \frac{dA(R)}{dR} \times \Delta R = 8\pi R\Delta R$$
$$= 8\pi(10 \text{ cm}) \times 0.1 \text{ cm}$$
$$= 8\pi \text{ cm}^2 = 25.1 \text{ cm}^2.$$

(c) The volume of the sphere is $V(R) = \frac{4}{3}\pi R^3$ and $\frac{dV(R)}{dR} = 4\pi R^2$,

$$\Delta V \approx \frac{dV(R)}{dR} \times \Delta R = 4\pi R^2\Delta R$$
$$= 4\pi(10 \text{ cm})^2 \times 0.1 \text{ cm}$$
$$= 40\pi \text{ cm}^3 = 125.6 \text{ cm}^3.$$

(d) We can use exact formulae to compute the changes of the quantities as follows:

(i)

$$\Delta C = C_{R=10.1\text{cm}} - C_{R=10\text{cm}}$$
$$= 2\pi(10.1 \text{ cm} - 10 \text{ cm})$$
$$= 0.2\pi \text{ cm}.$$

58

In this case the circumference computed using the approximation method in part (a) is the same as that computed by the exact method. We have perfect accuracy.

(ii)

$$\Delta A = A_{R=10.1\text{cm}} - A_{R=10\text{cm}}$$
$$= 4\pi \left(10.1^2 \text{ cm}^2 - 10^2 \text{ cm}^2\right)$$
$$= 8.04\pi \text{ cm}^2.$$

The error in using the approximation method is $\left|\frac{8.04\pi - 8\pi}{8.04\pi} \times 100\%\right| = 0.50\%$. We have high accuracy.

(iii)

$$\Delta V = V_{R=10.1\text{cm}} - V_{R=10\text{cm}}$$
$$= \frac{4}{3}\pi \left(10.1^3 \text{ cm}^3 - 10^3 \text{ cm}^3\right)$$
$$= 40.4\pi \text{ cm}^3.$$

The error in using the approximation method is $\left|\frac{40.4\pi - 40\pi}{40.4\pi} \times 100\%\right| = 1.0\%$. We have high accuracy.

---

14. We need to verify $\frac{df}{dx} = \left(\frac{dx}{df}\right)^{-1}$.

(i) Given $y = x^2$. First, we differentiate both sides with respect to x,

$$\frac{d}{dx}y = \frac{d}{dx}x^2$$
$$\therefore \frac{dy}{dx} = 2x.$$

Next, we differentiate both sides with respect to y and make use of the Chain Rule,

$$\frac{d}{dy}y = \frac{d}{dy}x^2 = \left(\frac{d}{dx}x^2\right) \times \left(\frac{dx}{dy}\right)$$
$$\Rightarrow 1 = 2x \times \frac{dx}{dy}$$
$$\therefore \frac{dx}{dy} = \frac{1}{2x}.$$

Hence $\frac{dx^2}{dx} = 2x = \left(\frac{dx}{dy}\right)^{-1}$.

(ii) Given that $y = \ln x$. Differentiate with respect to x,

$$\frac{d}{dx}y = \frac{d}{dx}\ln x$$
$$\therefore \frac{dy}{dx} = \frac{1}{x}.$$

Meanwhile, differentiating with respect to y gives

$$\frac{d}{dy}y = \frac{d}{dy}\ln x = \left(\frac{d}{dx}\ln x\right) \times \left(\frac{dx}{dy}\right)$$
$$\Rightarrow 1 = \frac{1}{x} \times \frac{dx}{dy}$$
$$\therefore \frac{dx}{dy} = x.$$

Hence $\frac{d\ln x}{dx} = \frac{1}{x} = \left(\frac{dx}{dy}\right)^{-1}$.

7.2 Solutions to Odd-Numbered Problems

1. We will discuss here the generic case $b \neq 0$, leaving the special case $b = 0$ to the reader.
(a) The first and second derivatives are

$$\frac{dy}{dx} = \frac{d}{dx}(ax^3 + bx^2 + cx + d)$$
$$= 3ax^2 + 2bx + c.$$
$$\frac{d^2y}{dx^2} = \frac{d}{dx}(3ax^2 + 2bx + c)$$
$$= 6ax + 2b.$$

(b) There are two possibilities,

- The function has no stationary points, that is, $\frac{dy}{dx} = 3ax^2 + 2bx + c = 0$ has no solutions. We obtain the discriminant condition $4b^2 - 4(3a)c < 0$ which simplifies to $b^2 < 3ac$.

- The function has a stationary point, but the second derivative test is inconclusive ($\frac{d^2y}{dx^2} = 0$), leading to the possibility that the point might be a point of inflexion rather than a turning point. This case therefore gives $\frac{dy}{dx} = 3ax^2 + 2bx + c = 0$ and $\frac{d^2y}{dx^2} = 6ax + 2b = 0$. Solving the simultaneous equations gives $x = -b/3a$ and $b^2 = 3ac$. However, for $b \neq 0$, we leave it for you to check that the first derivative test (see Exercise 8d) shows this point to be a turning point after all. (It is a point of inflexion for $b = 0$).

Therefore, we conclude that for the generic case $b \neq 0$, the condition for no turning points is $b^2 < 3ac$.

(c) To have two turning points, the equation $\frac{dy}{dx} = 3ax^2 + 2bx + c = 0$ should have two distinct real solutions, which implies the discriminant condition $(2b)^2 - 4(3a)c > 0$, that is $b^2 > 3ac$.

(d) Exactly one stationary point means $\frac{dy}{dx} = 0$ has an unique solution, which translates into $b^2 = 3ac$ and $x = -b/3a$. From part (a), we see that this also satisfies $\frac{d^2y}{dx^2} = 0$, but for $b \neq 0$ the first derivative test shows that this stationary point is a turning point rather than a point of inflexion.

_____

3. (a)

$$
\begin{aligned}
y' &= (\tan x)\frac{d(\sin^2 x)}{dx} + \frac{d(\tan x)}{dx}(\sin^2 x) \\
&= (\tan x)(2\sin x \cos x) + (\sec^2 x)(\sin^2 x) \\
&= 2\sin^2 x + \sin^2 x \sec^2 x \\
&= \sin^2 x(2 + \sec^2 x) = \sin^2 x(3 + \tan^2 x) \\
&= \tan^2 x(3\cos^2 x + \sin^2 x) \\
&= \tan^2 x(2\cos^2 x + 1) \\
&= \tan^2 x(\cos 2x + 2) \; .
\end{aligned}
$$

(b)

$$
\begin{aligned}
y' &= x\frac{d(e^{x^2})}{dx} + \frac{x}{x}e^{x^2} \\
&= x(2xe^{x^2}) + (1)e^{x^2} = (2x^2 + 1)e^{x^2} \; .
\end{aligned}
$$

(c)

$$
\begin{aligned}
y' &= \frac{(x^2 + 3x + 1)\frac{d(\sin x)}{dx} - \sin x\frac{d(x^2+3x+1)}{dx}}{(x^2 + 3x + 1)^2} \\
&= \frac{(x^2 + 3x + 1)\cos x - \sin x(2x + 3)}{(x^2 + 3x + 1)^2} \; .
\end{aligned}
$$

(d)

$$
\begin{aligned}
y' &= \cos(\ln(x^2 + 1))\frac{d(\ln(x^2 + 1))}{dx} \\
&= \cos(\ln(x^2 + 1)) \cdot \frac{1}{x^2 + 1} \cdot (2x) \\
&= \frac{2x\cos(\ln(x^2 + 1))}{x^2 + 1} \; .
\end{aligned}
$$

(e)

$$
\begin{aligned}
y' &= \frac{1}{\cos x^2} \cdot \frac{d(\cos x^2)}{dx} \\
&= \frac{1}{\cos x^2}(-\sin x^2)(2x) \\
&= -2x\tan x^2 \; .
\end{aligned}
$$

(f)

$$
\begin{aligned}
y' &= \frac{x^2 \frac{d(\ln 3x)^2}{dx} - (\ln 3x)^2(2x)}{x^4} \\
&= \frac{x^2(2\ln 3x) \cdot \frac{1}{3x} \cdot 3 - (\ln 3x)^2(2x)}{x^4} \\
&= \frac{2x\ln 3x - 2x(\ln 3x)^2}{x^4} \\
&= \frac{2\ln 3x(1 - \ln 3x)}{x^3} \; .
\end{aligned}
$$

5. Volume of sphere is $V = \frac{4\pi R^3}{3}$ with derivative $\frac{dV}{dR} = 4\pi R^2$. At $V = 50$ cm^3, we have $R = \sqrt[3]{\frac{150}{4\pi}}$. Also, we are given $\Delta V = 1$cm^3.

(a)

$$
\begin{aligned}
\Delta V &\approx \frac{dV}{dR}\Delta R \\
\Rightarrow \Delta R &\approx \frac{\Delta V}{dV/dR} = \frac{1}{(4\pi)^{1/3}(150)^{2/3}} \\
&= 0.015 \text{ cm.}
\end{aligned}
$$

(b) Direct calculations give,

$$
\begin{aligned}
V = 50 \text{ cm}^3 &\Rightarrow R = \sqrt[3]{\frac{150}{4\pi}} = 2.285 \text{ cm} \\
V = 51 \text{ cm}^3 &\Rightarrow R = \sqrt[3]{\frac{153}{4\pi}} = 2.300 \text{ cm} \\
\therefore \Delta R &= 0.015 \text{ cm.}
\end{aligned}
$$

_____

7. (a)

$$
\begin{aligned}
\frac{dy}{dx} &= \frac{d}{dx}3x^2(1 + x) + \frac{d}{dx}2\ln x \\
&= 6x(1 + x) + 3x^2 + \frac{2}{x} \\
&= 9x^2 + 6x + \frac{2}{x} \\
\left.\frac{d^2y}{dx^2}\right|_{x=1} &= 18x + 6 - \left.\frac{2}{x^2}\right|_{x=1} \\
&= 22.
\end{aligned}
$$

(b)

$$
\begin{aligned}
\frac{dy}{dx} &= \frac{d}{dx}\sin\ln x \\
&= \frac{\cos\ln x}{x} \\
\left.\frac{d^2y}{dx^2}\right|_{x=1} &= -\frac{\sin\ln x}{x^2} - \left.\frac{\cos\ln x}{x^2}\right|_{x=1} \\
&= -1 \; .
\end{aligned}
$$

60

(c)

$$\begin{aligned}
\frac{dy}{dx} &= \frac{d}{dx}\sqrt{2+3x^2}\\
&= \frac{3x}{\sqrt{2+3x^2}}\\
\frac{d^2y}{dx^2} &= \frac{3}{\sqrt{2+3x^2}} - \frac{9x^2}{(2+3x^2)^{3/2}}\\
\therefore\quad y''(1) &= \frac{6}{5\sqrt{5}} \approx 0.54 \ .
\end{aligned}$$

(d)

$$\begin{aligned}
\frac{dy}{dx} &= \frac{d}{dx}x\tan x\\
&= \tan x + x\sec^2 x\\
&= \tan x + x + x\tan^2 x\\
\frac{d^2y}{dx^2} &= 2\tan^2 x + 2x\tan x\sec^2 x\\
\therefore\quad y''(1) &= 17.52 \ .
\end{aligned}$$

(e)

$$\begin{aligned}
\frac{dy}{dx} &= \frac{d}{dx}x^2 e^x + \frac{d}{dx}\ln(1+x)\\
&= xe^x(2+x) + \frac{1}{1+x}\\
y''(x) &= e^x(2+4x+x^2) - \frac{1}{(1+x)^2}\\
y''(1) &= 7e - \frac{1}{4} \approx 18.78.
\end{aligned}$$

(f)

$$\begin{aligned}
\frac{dy}{dx} &= \frac{d}{dx}\frac{2+3x}{x+1}\\
&= \frac{3(x+1)-(2+3x)}{(x+1)^2}\\
&= \frac{1}{(x+1)^2}\\
\left.\frac{d^2y}{dx^2}\right|_{x=1} &= \frac{0(x+1)^2-2(x+1)}{(x+1)^4}\\
&= -\frac{2}{(x+1)^3} = -\frac{1}{4} \ .
\end{aligned}$$

---

9. (a) Since the exponential function $e^{-t} > 0$ for all t, $x(t) = 0 \Rightarrow \sin(12t) = 0$. Thus, $12t = n\pi$ for $n = 0, 1, 2....$ So the particle first returns to the origin when $n = 1$, or $t = \pi/12$.

(b)

$$\begin{aligned}
v_x(t) &= \frac{dx}{dt}\\
&= -2e^{-t}\sin(12t) + 24e^{-t}\cos(12t)\\
&= 2e^{-t}\big[-\sin(12t) + 12\cos(12t)\big]\\
\left.v_x\right|_{t=\frac{\pi}{12}} &= 2e^{-\pi/12}\big[-\sin\pi + 12\cos\pi\big]\\
&= -24e^{-\pi/12} \approx -18.5 \ .
\end{aligned}$$

(c) For the velocity to become zero,

$$\begin{aligned}
0 &= 2e^{-t}\big[-\sin(12t) + 12\cos(12t)\big]\\
\tan(12t) &= 12\\
t &= \frac{\tan^{-1}12}{12} \approx 0.124 \ .
\end{aligned}$$

(d)

$$\begin{aligned}
a_x(t) &= \frac{dv}{dt}\\
&= -2e^{-t}\big[-\sin(12t) + 12\cos(12t)\big]\\
&\quad + 2e^{-t}\big[-12\cos(12t) - 144\sin(12t)\big]\\
&\Rightarrow \left.a_x\right|_{t=0.124} \approx -255 \ .
\end{aligned}$$

(e) Setting $a(t) = 0$ gives

$$\begin{aligned}
0 &= 24\cos(12t) + 143\sin(12t)\\
\tan(12t) &= -\frac{24}{143} \Rightarrow 12t = \pi - \tan^{-1}\frac{24}{143}\\
\therefore t &\approx 0.248 \ .
\end{aligned}$$

(f)

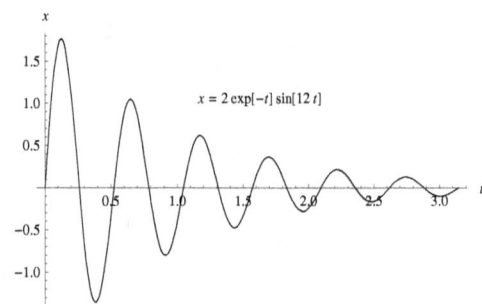

Figure 7.8: Plot of displacement function for Problem 9.

---

11. Suppose the length and the width of the rectangular section are l and w. The perimeter gives the constraint $L = 2(l + w)$. So the area is $A = l \times w = \frac{Ll}{2} - l^2$.

(a) To maximize the enclosed area, we have to ensure $dA/dl = 0$ and $d^2A/dl^2 < 0$.

$$\frac{dA}{dl} = \frac{L}{2} - 2l = 0$$
$$\Rightarrow l = \frac{L}{4}$$

Furthermore, $\frac{d^2A}{dl^2} = -2 < 0$, so the solution $l = L/4, w = L/4$ is a local maximum. The dimension of the rectangle with maximum area is $L/4 \times L/4$, which is a square.

We can also obtain the maximum area by completing the square

$$A = -l^2 + \frac{Ll}{2} = -\left(l - L/4\right)^2 + \frac{L^2}{16} \le \frac{L^2}{16}.$$

The area is maximum when the term in bracket vanishes, that is $(l - L/4)^2 = 0 \Rightarrow l = L/4$.

(b) From part (a), we see that in the limit $l \to 0$, the area $A = -(l - L/4)^2 + L^2/16$ is decreasing and approaches zero. Thus the minimum area is when we have a finite width but infinitesimal length, or vice-versa.

_____

13. Let us sketch the diagram of the triangle and inscribed squared.

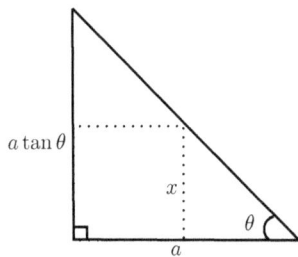

Figure 7.9: Sketch for Problem 13.

(a) The area of the triangle is $A = \frac{1}{2}a^2 \tan\theta$. Suppose we let the side of the square be x, we have $\tan\theta = \frac{x}{a-x} \Rightarrow x = \frac{a\tan\theta}{1+\tan\theta}$. Thus the area of the square is

$$S = x^2 = \frac{a^2 \tan^2\theta}{(1+\tan\theta)^2} = \frac{2A\tan\theta}{(1+\tan\theta)^2}.$$

(b) Differentiating,

$$\frac{dS}{d\theta} = 2A\sec^2\theta\frac{(1-\tan\theta)}{(1+\tan\theta)^3} = 0$$
$$\tan\theta = 1 \Rightarrow \theta = \pi/4.$$

The second derivative is

$$\left.\frac{d^2S}{d\theta^2}\right|_{\theta=\pi/4} = \left.\frac{4A(-2+\sin(2\theta))}{(\cos\theta+\sin\theta)^4}\right|_{\theta=\pi/4}$$
$$= -A < 0.$$

Thus, the maximum area of the square is given by $S_{max} = \frac{a^2\tan^2\pi/4}{(1+\tan\pi/4)^2} = \frac{A}{2}$.

(c) First, we note that the area of the inscribed square goes to zero as $\theta \to 0$. We also note from the figure that if one acute angle is labelled θ, the other is $\pi/2 - \theta$; therefore $S(\theta)$ must be the same as $S(\pi/2-\theta)$ (you can check this from the explicit expression for S). Therefore S also goes to zero as $\theta \to \pi/2$. So, if $S(\theta)$ has only one maximum, it must be at the symmetrical midpoint $\theta = \pi/4$. Let us check this by making a change of variables. Replace θ by $\pi/4 - x$ in $S(\theta)$ and simplify to find $S = \cos(2x)/2$, which decreases when x deviates from zero. So $\theta = \pi/4$ is indeed the location of the unique maximum.

_____

Did You Know?

Ants can find the shortest path from their nest to a food source.

How do they do it? Calculus? Not really! They arrive at the path by a process of "self-organisation".

Self-organisation is one of the key concepts in the study of complex systems.

The book *Simplicity in Complexity: An Introduction to Complex Systems*, by R. Parwani, discusses various concepts and tools used in complexity studies.

View sample pages at www.simplicitysg.net/books/sic

Chapter 8

Integral Calculus

8.1   Solutions o Exercises

1. *Recall: For a curve C with $y(x) \geq 0$, the area between the curve and x-axis is given by $\int_a^b y(x)\ dx$.*

(a) Given the curve $y = \sin x$, where $x \in [0, \pi/2]$,

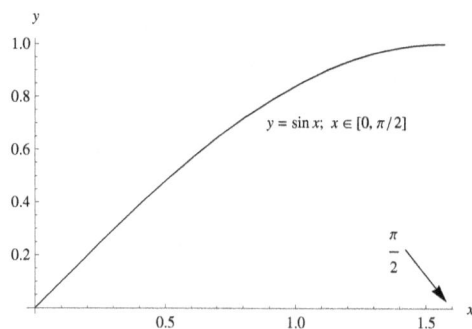

Figure 8.1: Plot of $y(x) = \sin x$.

Area between $y = \sin x$ and the x-axis is,

$$
\begin{aligned}
A &= \int_0^{\pi/2} \sin x\ dx = [-\cos x]_0^{\pi/2} \\
&= -\cos \frac{\pi}{2} - [-\cos 0] \\
&= 0 + 1 = 1.
\end{aligned}
$$

(b) Given the curve $y = e^x$, where $x \in (-\infty, 0]$,

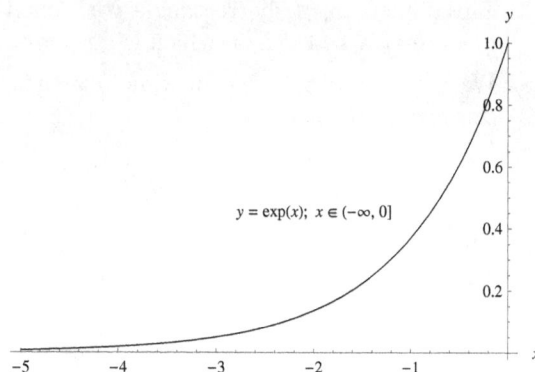

Figure 8.2: Plot of $y(x) = \exp(x)$.

Area between $y = e^x$ and the x-axis is,

$$
\begin{aligned}
A &= \int_{-\infty}^0 e^x\ dx = [e^x]_{-\infty}^0 \\
&= e^0 - e^{-\infty} \\
&= 1 - 0 = 1.
\end{aligned}
$$

(c) Given the curve $y = \frac{1}{x}$, where $x \in [1, 2]$,

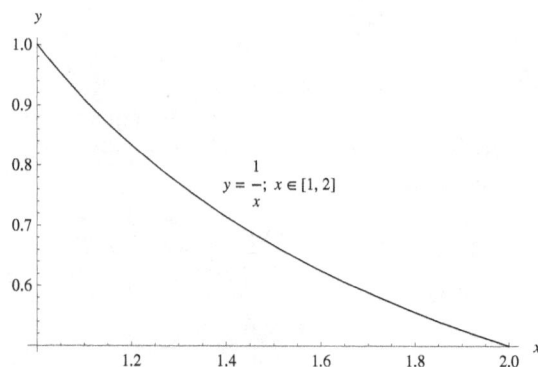

Figure 8.3: Plot of $y(x) = \frac{1}{x}$.

Area between $y = \frac{1}{x}$ and the x-axis is,

$$
\begin{aligned}
A &= \int_1^2 \frac{1}{x}\, dx = [\ln x]_1^2 \\
&= \ln 2 - \ln 1 \\
&= \ln \frac{2}{1} = \ln 2.
\end{aligned}
$$

(d) Given the curve $y = \frac{1}{x^2}$, where $x \in [1, \infty)$,

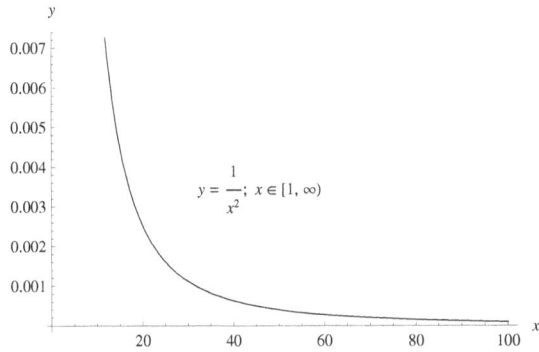

Figure 8.4: Plot of $y(x) = \frac{1}{x^2}$.

Area between $y = \frac{1}{x^2}$ and the x-axis is,

$$
\begin{aligned}
A &= \int_1^\infty \frac{1}{x^2}\, dx = \left[-\frac{1}{x}\right]_1^\infty \\
&= -\frac{1}{\infty} - \left(-\frac{1}{1}\right) \\
&= 0 + 1 = 1.
\end{aligned}
$$

(e) Given the curve $y = \sqrt{x}$, where $x \in [0, 1]$,

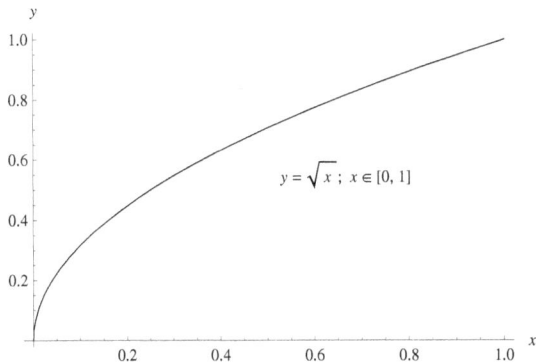

Figure 8.5: Plot of $y(x) = \sqrt{x}$.

Area between $y = \sqrt{x}$ and the x-axis is,

$$
\begin{aligned}
A &= \int_0^1 \sqrt{x}\, dx = \left[\frac{2}{3}x^{3/2}\right]_0^1 \\
&= \frac{2}{3}1^{3/2} - 0 \\
&= \frac{2}{3}.
\end{aligned}
$$

(f) Given the curve $y = x^2$, where $x \in [0, 1]$,

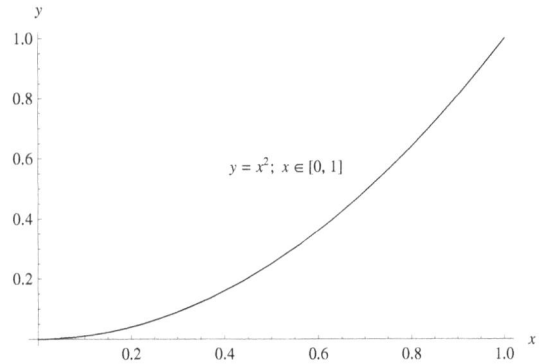

Figure 8.6: Plot of $y(x) = x^2$.

Area between $y = x^2$ and the x-axis is,

$$
\begin{aligned}
A &= \int_0^1 x^2\, dx = \left[\frac{x^3}{3}\right]_0^1 \\
&= \frac{1^3}{3} - 0 \\
&= \frac{1}{3}.
\end{aligned}
$$

2. (a)

$$
\begin{aligned}
I &= \int_0^{2\pi} \sin x\, dx \\
&= [-\cos x]_0^{2\pi} \\
&= -\cos(2\pi) - [-\cos 0] \\
&= -1 - (-1) = 0.
\end{aligned}
$$

(b) To find the area, we first plot the curve $y = \sin x$ for $x \in [0, 2\pi]$.

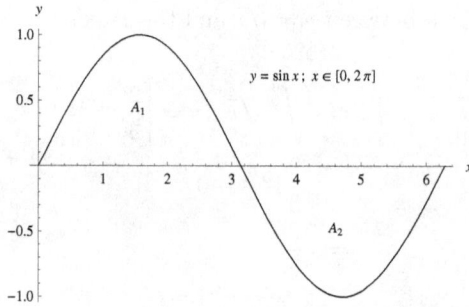

Figure 8.7: Plot of $y(x) = \sin x$.

From the graph, we see that there is a segment A_2 located below the x-axis, and the integration along that stretch would give a negative value. Therefore to get the area between the x-axis and the curve, we need to divide the integration into two parts and add the magnitudes. We have

$$
\begin{aligned}
A_1 &= \int_0^\pi \sin x \, dx \\
&= [-\cos x]_0^\pi = [-\cos \pi] - [-\cos 0] \\
&= 1 - (-1) = 2. \\
A_2 &= \left| \int_\pi^{2\pi} \sin x \, dx \right| \\
&= \left| [-\cos x]_\pi^{2\pi} \right| \\
&= \left| [-\cos 2\pi] - [-\cos \pi] \right| \\
&= \left| -1 - (1) \right| = 2.
\end{aligned}
$$

Therefore $A = A_1 + A_2 = 2 + |-2| = 4$.

(c) It is clear that in part (a) half of the integral contributes a negative amount which exactly cancels the positive contribution from the other half.

3. (a) *Note:* $\int (a+bx)^n \, dx = \frac{(a+bx)^{n+1}}{b(n+1)} + C$, and $\int \frac{1}{a+bx} \, dx = \frac{1}{b} \ln(a+bx) + C$, where C is a constant of integration.

$$
\begin{aligned}
&\int \left(3x^2(1+x) - \frac{2}{x} \right) dx \\
&= \int \left(3x^2 + 3x^3 - \frac{2}{x} \right) dx \\
&= \frac{3x^{2+1}}{(2+1)} + \frac{3x^{3+1}}{(3+1)} - 2\ln x + C \\
&= x^3 + \frac{3x^4}{4} - 2\ln x + C.
\end{aligned}
$$

(b)

$$
\begin{aligned}
&\int \sqrt{2+3x} \, dx \\
&= \frac{(2+3x)^{1/2+1}}{3(1/2+1)} + C \\
&= \frac{2(2+3x)^{3/2}}{9} + C.
\end{aligned}
$$

(c) *Note:* $\int \sin(a+bx) \, dx = -\frac{1}{b}\cos(a+bx) + C$, and $\int \sec^2(a+bx) \, dx = \frac{1}{b}\tan(a+bx) + C$.

$$
\begin{aligned}
&\int \left(2\sin 2x - 3\sec^2 x \right) dx \\
&= 2\left(-\frac{1}{2}\cos 2x \right) - 3(\tan x) + C \\
&= -\cos 2x - 3\tan x + C.
\end{aligned}
$$

(d) *Note:* $\int \frac{1}{a+bx} \, dx = \frac{1}{b}\ln(a+bx) + C$.

$$
\begin{aligned}
&\int \frac{2}{5+3x} \, dx \\
&= 2\left(\frac{1}{3}\ln(5+3x) \right) + C \\
&= \frac{2\ln(5+3x)}{3} + C.
\end{aligned}
$$

(e) *Note:* $\int e^{a+bx} \, dx = \frac{1}{b}e^{a+bx} + C$.

$$
\begin{aligned}
&\int \left(2e^{3x} - (1+x)^3 \right) dx \\
&= 2\left(\frac{e^{3x}}{3} \right) - \frac{(1+x)^{3+1}}{(3+1)} + C \\
&= \frac{2e^{3x}}{3} - \frac{(1+x)^4}{4} + C.
\end{aligned}
$$

(f) *Note:* $\sin 2x = 2\sin x \cos x$.

$$
\begin{aligned}
&\int 7\sin 2x \cos 2x \, dx \\
&= \frac{7}{2} \int \sin 4x \, dx \\
&= \frac{7}{2} \left[-\frac{\cos 4x}{4} \right] + C \\
&= -\frac{7\cos 4x}{8} + C.
\end{aligned}
$$

(g)

$$
\begin{aligned}
&\int \left(\frac{1}{e^{2x}} - \frac{1}{\sqrt{2+3x}} \right) dx \\
&= \int \left(e^{-2x} - (2+3x)^{-1/2} \right) dx \\
&= \frac{e^{-2x}}{-2} - \left[\frac{1}{3} \times \frac{(2+3x)^{-1/2+1}}{(-1/2+1)} \right] + C \\
&= -\frac{1}{2e^{2x}} - \frac{2\sqrt{2+3x}}{3} + C.
\end{aligned}
$$

4. *Note: Except for some differences in sign, the integrations here are similar to those done in the previous question. The reader is therefore advised to check details there.*

(a)

$$\int_1^2 \left(3x^2(1+x) + \frac{2}{x} \right) dx$$

$$= \left[x^3 + \frac{3x^4}{4} + 2\ln x \right]_1^2$$

$$= \left[2^3 + \frac{3(2^4)}{4} + 2\ln 2 \right] - \left[1 + \frac{3}{4} + 2\ln 1 \right]$$

$$= 7 + \frac{45}{4} + 2\ln 2 \approx 19.64 .$$

(b)

$$\int_0^1 \sqrt{2+3x}\, dx$$

$$= \left[\frac{2(2+3x)^{3/2}}{9} \right]_0^1$$

$$= \left[\frac{2(2+3)^{3/2}}{9} \right] - \left[\frac{2(2)^{3/2}}{9} \right]$$

$$= \frac{2}{9}(5^{3/2} - 2^{3/2}) \approx 1.86 .$$

(c)

$$\int_{-1}^1 \frac{2}{5+3x}\, dx$$

$$= \left[\frac{2\ln(5+3x)}{3} \right]_{-1}^1$$

$$= \left[\frac{2\ln(5+3)}{3} \right] - \left[\frac{2\ln(5-3)}{3} \right]$$

$$= \left[\frac{2\ln 8}{3} \right] - \left[\frac{2\ln 2}{3} \right] = \frac{2}{3}\ln\frac{8}{2}$$

$$= \frac{2}{3}\ln 4 \approx 0.92 .$$

(d)

$$\int_0^{\pi/4} \left(2\sin 2x + 3\sec^2 x \right) dx$$

$$= [-\cos 2x + 3\tan x]_0^{\pi/4}$$

$$= [-\cos 2(\pi/4) + 3\tan(\pi/4)]$$

$$\quad - [-\cos 2(0) + 3\tan 0]$$

$$= [-0 + 3(1)] - [-1 - 3(0)]$$

$$= 3 + 1 = 4 .$$

(e)

$$\int_0^1 \left(2e^{3x} + (1+x)^3 \right) dx$$

$$= \left[\frac{2e^{3x}}{3} + \frac{(1+x)^4}{4} \right]_0^1$$

$$= \left[\frac{2e^3}{3} + \frac{(1+1)^4}{4} \right]$$

$$\quad - \left[\frac{2e^0}{3} + \frac{(1+0)^4}{4} \right]$$

$$= \frac{2}{3}(e^3 - e^0) + \frac{15}{4}$$

$$\approx 16.47 .$$

(f)

$$\int_0^2 \left(\frac{1}{e^{2x}} + \frac{1}{\sqrt{2+3x}} \right) dx$$

$$= \left[-\frac{1}{2e^{2x}} + \frac{2\sqrt{(2+3x)}}{3} \right]_0^2$$

$$= \left[-\frac{1}{2e^{2(2)}} + \frac{2\sqrt{(2+3(2))}}{3} \right]$$

$$\quad - \left[-\frac{1}{2e^0} + \frac{2\sqrt{(2+3(0))}}{3} \right]$$

$$= -\frac{1}{2e^4} + \frac{2\sqrt{8}}{3} + \frac{1}{2} - \frac{2\sqrt{2}}{3}$$

$$= \frac{1}{2}(1 - e^{-4}) + \frac{2}{3}(2\sqrt{2} - \sqrt{2}) \approx 1.43 .$$

(g)

$$\int_0^\pi \sin 2x \cos 2x \, dx$$

$$= \frac{1}{7}\left[-\frac{7\cos 4x}{8} \right]_0^\pi$$

$$= \left[-\frac{\cos 4\pi}{8} \right] - \left[-\frac{\cos 4(0)}{8} \right]$$

$$= -\frac{1}{8} - \left(-\frac{1}{8} \right)$$

$$= 0 .$$

5. If the slope of the normal is $\frac{x}{1-x}$, then the slope of the tangent to the curve at the same point will be $\frac{dy}{dx} = -\left(\frac{1-x}{x} \right)$. We can integrate the last expression with respect to x to obtain the equation of the curve.

$$\frac{dy}{dx} = -\frac{1-x}{x}$$

$$\Rightarrow y(x) = \int -\frac{1-x}{x}\, dx = \int \left(-\frac{1}{x} + 1 \right) dx$$

$$= -\ln x + x + C.$$

Since the curve passes through the point $(1, 2)$, $C = 2 + \ln 1 - 1 = 1$. Hence, the equation of the curve is

$$y(x) = -\ln x + x + 1.$$

---

6. Given that $\frac{dR}{dt} = 0.5$ cm/s and $R = 10$ cm at $t = 0$.

$$\frac{dR}{dt} = 0.5$$

$$\Rightarrow R(t) = \int 0.5 \, dt = 0.5t + C.$$

We can fix the constant C by the initial condition, $C = 10 - 0.5(0) = 10$ cm. So

$$R(t) = 0.5t + 10 .$$
$$R_{(t=5s)} = 0.5(5) + 10 = 12.5 \text{ cm}.$$

---

7. Since the maximum value for $\sin^2 x$ is 1, then thinking in terms of the area under the curve,

$$\int_0^{2\pi} \sin^2 x \, dx < \int_0^{2\pi} 1 \, dx = [x]_0^{2\pi} = 2\pi.$$

Of course one can evaluate the integral exactly,

$$\int_0^{2\pi} \sin^2 x \, dx = \int_0^{2\pi} \frac{1 - \cos 2x}{2} \, dx$$
$$= \frac{1}{2} \left[x - \frac{\sin 2x}{2} \right]_0^{2\pi}$$
$$= \frac{1}{2} \left[2\pi - \frac{\sin 2(2\pi)}{2} \right] - \frac{1}{2} \left[0 - \frac{\sin 2(0)}{2} \right]$$
$$= \pi < 2\pi.$$

---

8. First we find the turning points by setting the first derivative to zero.

$$\frac{dy}{dx} = \frac{d}{dx} [x(x-1)(x-3)]$$
$$= \frac{d}{dx} [x^3 - 4x^2 + 3x]$$
$$= 3x^2 - 8x + 3 .$$

Set $\frac{dy}{dx} = 0$

$$\Rightarrow x = \frac{8 \pm \sqrt{64 - 4(3)(3)}}{2(3)}$$
$$= \frac{4 \pm \sqrt{7}}{3}.$$

We have the turning points $x_1 = \frac{4-\sqrt{7}}{3}$ and $x_2 = \frac{4+\sqrt{7}}{3}$. Next, we sketch the curve:

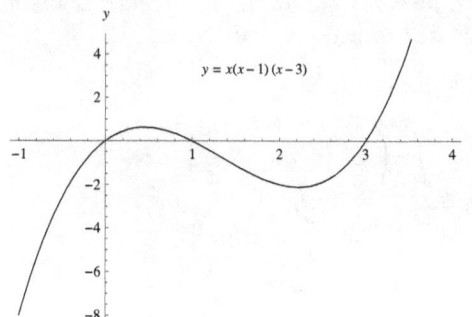

Figure 8.8: Plot of $y(x) = x(x-1)(x-3)$.

The area between consecutive turning points is (note the modulus over the negative segment)

$$A = \int_{x_1}^1 y \, dx + \left| \int_1^{x_2} y \, dx \right|$$
$$= \int_{x_1}^1 y \, dx + \int_{x_2}^1 y \, dx$$
$$= \int_{x_1}^1 x^3 - 4x^2 + 3x \, dx$$
$$\quad + \int_{x_2}^1 x^3 - 4x^2 + 3x \, dx$$
$$= \left[\frac{x^4}{4} - \frac{4x^3}{3} + \frac{3x^2}{2} \right]_{x_1}^1$$
$$\quad + \left[\frac{x^4}{4} - \frac{4x^3}{3} + \frac{3x^2}{2} \right]_{x_2}^1$$
$$= - \left[\frac{(x_1^4 + x_2^4)}{4} - \frac{4(x_1^3 + x_2^3)}{3} \right.$$
$$\left. + \frac{3(x_1^2 + x_2^2)}{2} \right] + 2 \left[\frac{1}{4} - \frac{4}{3} + \frac{3}{2} \right].$$

We can evaluate each term in the above equation exactly as follows:

$$x_1^2 + x_2^2 = \left(\frac{4 - \sqrt{7}}{3} \right)^2 + \left(\frac{4 + \sqrt{7}}{3} \right)^2$$
$$= \frac{23 - 8\sqrt{7}}{9} + \frac{23 + 8\sqrt{7}}{9} = \frac{46}{9};$$

$$x_1^3 + x_2^3 = \left(\frac{4-\sqrt{7}}{3}\right)^3 + \left(\frac{4+\sqrt{7}}{3}\right)^3$$

$$= \frac{(23-8\sqrt{7})(4-\sqrt{7})}{27}$$

$$+\frac{(23+8\sqrt{7})(4+\sqrt{7})}{27}$$

$$= \frac{148-55\sqrt{7}}{27} + \frac{148+55\sqrt{7}}{27} = \frac{296}{27};$$

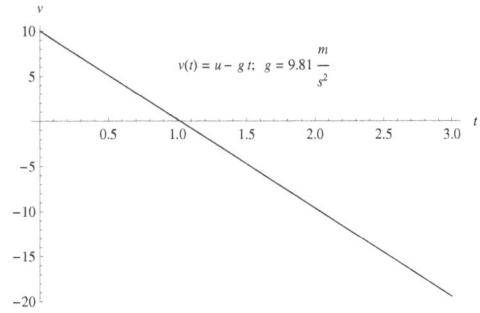

Figure 8.9: Plot of $v(t) = u - gt$. As an example, we have set $u = 10$ m/s and $g = 9.81$ m s^{-2}.

$$x_1^4 + x_2^4 = \left(\frac{4-\sqrt{7}}{3}\right)^4 + \left(\frac{4+\sqrt{7}}{3}\right)^4$$

$$= \frac{(148-55\sqrt{7})(4-\sqrt{7})}{81}$$

$$+\frac{(148+55\sqrt{7})(4+\sqrt{7})}{81}$$

$$= \frac{977-368\sqrt{7}}{81} + \frac{977+368\sqrt{7}}{81} = \frac{1954}{81}.$$

Hence the area is

$$A = -\left[\frac{1954}{4(81)} - \frac{4(296)}{3(27)} + \frac{3(46)}{2(9)}\right]$$

$$+2\left[\frac{3-16+18}{12}\right]$$

$$= -\left[\frac{1954}{324} - \frac{1184}{81} + \frac{138}{18}\right] + \frac{5}{6}$$

$$= \frac{-1954 + 1184(4) - 138(18) + 5(54)}{324}$$

$$= \frac{568}{328} = \frac{142}{81}.$$

(b) Similarly, the displacement vector can be obtained by integrating the velocity vector with respect to t,

$$y(t) = \int (u - gt)\, dt = ut - \frac{1}{2}gt^2 + C_2.$$

By using the initial value, we can fix the constant $C_2 = 0 - u(0) + \frac{1}{2}g(0)^2 = 0$. So we have the displacement $y(t) = ut - \frac{1}{2}gt^2$ sketched below.

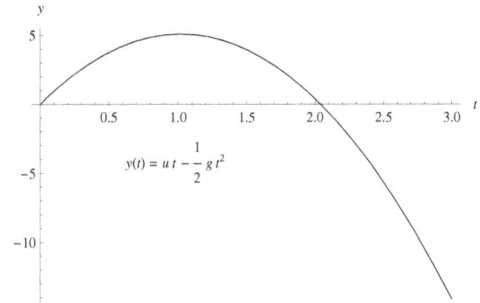

Figure 8.10: Plot of $y(t) = ut - \frac{1}{2}gt^2$. We set $u = 10$ m/s and $g = 9.81$ m s^{-2}.

(c) When it reaches maximum height, it is at a turning point where $v(t) = \dot{y} = 0$. Thus we have

$$v(t) = u - gt = 0$$

$$\Rightarrow t = \frac{u}{g}.$$

(d) We can substitute $t = \frac{u}{g}$ in $y(t)$ to obtain the maximum height,

$$y_{\max} = \left[ut - \frac{1}{2}gt^2\right]_{t=u/g}$$

$$= u\left(\frac{u}{g}\right) - \frac{1}{2}g\left(\frac{u}{g}\right)^2$$

$$= \frac{u^2}{2g}.$$

9. (a) Velocity is obtained by integrating the acceleration vector with respect to time,

$$v(t) = \int a\, dt = \int -g\, dt$$

$$= -gt + C_1.$$

By using the initial value, we can fix the constant $C_1 = u + g(0) = u$. So the velocity is $v(t) = u - gt$. See sketch below.

(e) When it returns to $y(t) = 0$,

$$0 = ut - \frac{1}{2}gt^2 = t(u - \frac{1}{2}gt)$$
$$\therefore t = 0 \text{ or } \frac{2u}{g}.$$

So the time taken for the particle to return to $y = 0$ is $t = \frac{2u}{g}$ s.

(f) The velocity of particle of particle when it returns to $y = 0$ is

$$v(t) = u - g \times \left(\frac{2u}{g}\right) = -u.$$

(g) This equation describes the vertical motion of a particle in a constant gravitational field.

10. Given that $\int_1^6 f(x)\,dx = 20$ and $\int_1^8 f(x)\,dx = 15$, then

(a)
$$\int_1^8 f(x)\,dx = \int_1^6 f(x)\,dx + \int_6^8 f(x)\,dx$$
$$\therefore \int_6^8 f(x)\,dx = \int_1^8 f(x)\,dx - \int_1^6 f(x)\,dx$$
$$= 15 - 20 = -5.$$

(b) Since $\int_1^6 f(x)\,dx = 20 > 0$, the function $f(x)$ must be positive for some $x \in [1,6]$.

(c) Since $\int_6^8 f(x)\,dx = -5 < 0$, the function $f(x)$ has to be negative for some $x \in [6,8]$.

8.2 Solutions to Odd-Numbered Problems

1. (a)
$$\int_0^{\pi/2} \sin^2 x\,dx = \frac{1}{2}\int_0^{\pi/2}(1-\cos 2x)\,dx$$
$$= \frac{1}{2}\left[x - \frac{\sin 2x}{2}\right]_0^{\pi/2} = \frac{\pi}{4}.$$

(b)
$$\int_0^{\pi/2}(\cos^2 x - \sin^2 x)\,dx = \int_0^{\pi/2}\cos 2x\,dx$$
$$= \left[\frac{\sin 2x}{2}\right]_0^{\pi/2} = 0.$$

(c)
$$\int_0^{\pi/4}\sec^2 x\,dx = \left[\tan x\right]_0^{\pi/4} = 1.$$

(d)
$$\int_0^1 \frac{1}{\sqrt{x}}\,dx = \left[2\sqrt{x}\right]_0^1 = 2.$$

3. (a) We can re-write the integrand using partial fractions and perform the integrations directly,

$$\int \frac{2x+1}{x^2+3x+2}\,dx$$
$$= \int \frac{3}{(x+2)}\,dx - \int \frac{1}{(x+1)}\,dx$$
$$= 3\ln(x+2) - \ln(x+1) + C.$$

(b) Performing a long division and a partial fraction decomposition, we have

$$\int \frac{5x^2+2x+1}{x^2+3x+2}\,dx$$
$$= \int 5\,dx - \int \frac{17}{(x+2)}\,dx + \int \frac{4}{(x+1)}\,dx$$
$$= 5x - 17\ln(x+2) + 4\ln(x+1) + C.$$

(c) As in part (b) above,

$$\int \frac{5x^2+2x+1}{x+1}\,dx$$
$$= \int (5x-3)\,dx + \int \frac{4}{(x+1)}\,dx$$
$$= \frac{5}{2}x^2 - 3x + 4\ln(x+1) + C.$$

(d) As before,

$$\int \frac{5x^2+2x+1}{x^2+2x+1}\,dx$$
$$= \int 5\,dx - \int \frac{8}{(x+1)}\,dx + \int \frac{4}{(x+1)^2}\,dx$$
$$= 5x - 8\ln(x+1) - \frac{4}{(x+1)} + C.$$

(e) As above,

$$\int \frac{x^4+1}{x^3+x}\,dx$$
$$= \int x\,dx + \int \frac{1}{x}\,dx - \int \frac{2x}{x^2+1}\,dx$$
$$= \frac{1}{2}x^2 + \ln x - \ln(x^2+1) + C.$$

5. (a)

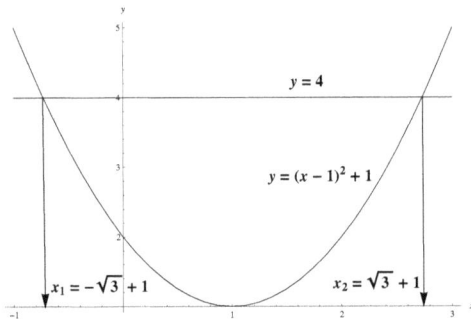

Figure 8.11: Plot of $y(x) = (x-1)^2 + 1$ and $y = 4$ for Problem 5a.

The two curves intersect at $x_{1,2} = \pm\sqrt{3} + 1$, with corresponding $y = 4$. The area bounded by these two curves is given by the area of the rectangle, $4(x_2 - x_1)$, minus the area below the parabola:

$$
\begin{aligned}
A &= 4(2\sqrt{3}) - \int_{-\sqrt{3}+1}^{\sqrt{3}+1} \left((x-1)^2 + 1\right) dx \\
&= 8\sqrt{3} - \left[\frac{(x-1)^3}{3} + x\right]_{-\sqrt{3}+1}^{\sqrt{3}+1} \\
&= 8\sqrt{3} - \left[(2\sqrt{3}+1) - (-2\sqrt{3}+1)\right] \\
&= 4\sqrt{3}.
\end{aligned}
$$

(b)

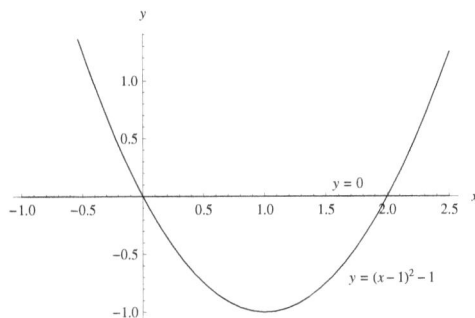

Figure 8.12: Plot of $y(x) = (x-1)^2 - 1$ and $y = 0$.

The two curves intersect at $x = 0, 2$. The

integration involved is

$$
\begin{aligned}
I &= \int_0^2 \left((x-1)^2 - 1\right) dx \\
&= \left[\frac{(x-1)^3}{3} - x\right]_0^2 \\
&= \left[(\frac{1}{3} - 2) - (-\frac{1}{3})\right] \\
&= -\frac{4}{3}.
\end{aligned}
$$

The area bounded by these two curves is given by $A = |I| = \frac{4}{3}$.

(c)

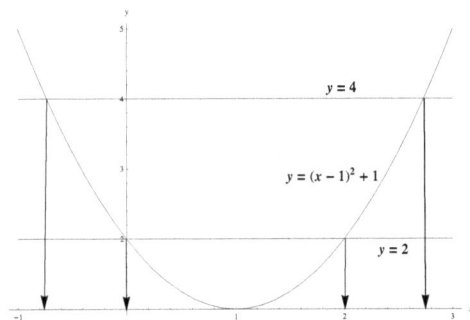

Figure 8.13: Plot of $y(x) = (x-1)^2 + 1$, $y = 2$ and $y = 4$ for Problem 5c.

From the figure, we see that the area needed is part of the answer to Problem (5a). We need to subtract out the smaller area below $y = 2$, which can be determined in a similar way. The result is

$$
\begin{aligned}
A &= 4\sqrt{3} - \left(4 - \int_0^2 \left((x-1)^2 + 1\right) dx\right) \\
&= 4\sqrt{3} - 4 + \left[\frac{(x-1)^3}{3} + x\right]_0^2 \\
&= \left(4\sqrt{3} - \frac{4}{3}\right).
\end{aligned}
$$

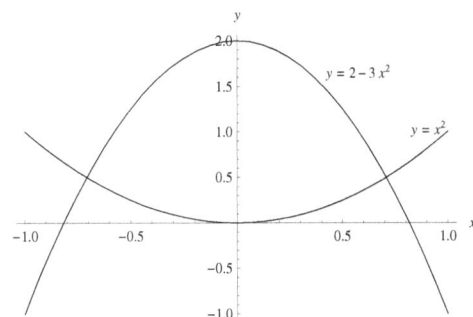

Figure 8.14: Plot of $y = 2 - 3x^2$ and $y = x^2$.

(d) From the figure above, the intersection of the two curves is at $x = \pm\frac{1}{\sqrt{2}}$. The area bounded is thus,

$$
\begin{aligned}
A &= \int_{-\frac{1}{\sqrt{2}}}^{\frac{1}{\sqrt{2}}} \left((2 - 3x^2) - x^2 \right) \\
&= \left[2x - \frac{4}{3}x^3 \right]_{-\frac{1}{\sqrt{2}}}^{\frac{1}{\sqrt{2}}} \\
&= \frac{8}{3\sqrt{2}} = \frac{4\sqrt{2}}{3}.
\end{aligned}
$$

(e)

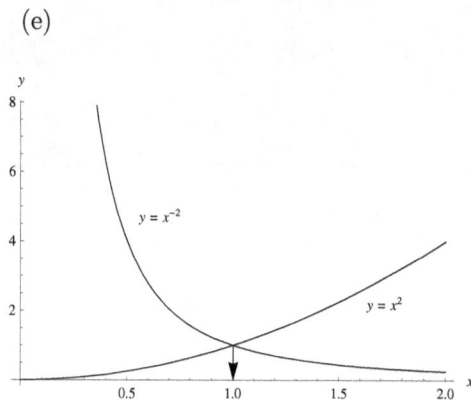

Figure 8.15: Plot of $y = x^2$ and its reciprocal $y = 1/x^2$.

The intersection of the two curves for $x \geq 0$ is at $x = 1$. The area bounded is thus (see figure),

$$
\begin{aligned}
A &= \int_0^1 x^2\,dx + \int_1^\infty \frac{dx}{x^2} \\
&= \left[\frac{x^3}{3} \right]_0^1 - \left[\frac{1}{x} \right]_1^\infty = \frac{4}{3}.
\end{aligned}
$$

---

7. (a)

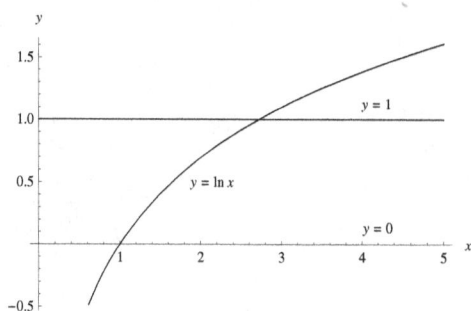

Figure 8.16: Plot of $y = \ln x$, $y = 0$ and $y = 1$.

Using integration with respect to y axis,

$$
\begin{aligned}
A &= \int_0^1 x\,dy = \int_0^1 e^y\,dy \\
&= \left[e^y \right]_0^1 = (e - 1) \approx 1.72.
\end{aligned}
$$

(b)

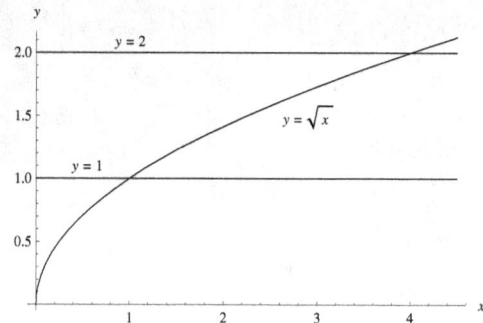

Figure 8.17: Plot of $y = \sqrt{x}$, $y = 1$ and $y = 2$.

Using integration with respect to y axis,

$$
\begin{aligned}
A &= \int_1^2 x\,dy = \int_1^2 y^2\,dy \\
&= \left[\frac{y^3}{3} \right]_1^2 = \frac{7}{3}.
\end{aligned}
$$

(c)

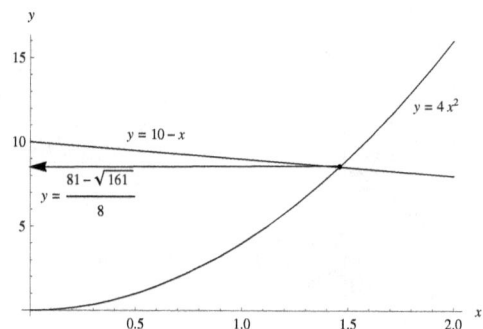

Figure 8.18: Plot of $y = 4x^2$ and $y = 10 - x$.

The intersection of the two curves for $x \geq 0$ is at $(x, y) = \left(\frac{-1+\sqrt{161}}{8}, \frac{81-\sqrt{161}}{8} \right)$. The area bounded is thus (see figure),

$$
\begin{aligned}
A &= \int_0^{\frac{81-\sqrt{161}}{8}} x_1 \, dy + \int_{\frac{81-\sqrt{161}}{8}}^{10} x_2 \, dy \\
&= \int_0^{\frac{81-\sqrt{161}}{8}} \frac{\sqrt{y}}{2} dy + \int_{\frac{81-\sqrt{161}}{8}}^{10} (10-y) dy \\
&= \left[\frac{y^{3/2}}{3} \right]_0^{\frac{81-\sqrt{161}}{8}} + \left[10y - \frac{y^2}{2} \right]_{\frac{81-\sqrt{161}}{8}}^{10} \\
&= 9.38.
\end{aligned}
$$

---

9. We integrate $y = f(x)$ with the boundary condition $f'(x) = 0$ at $y = 1$.

(a)

$$
\begin{aligned}
f(x) &= \int f'(x) dx \\
&= \int \left((1+x) - \frac{4}{x+1} \right) dx \\
&= \frac{1}{2}(1+x)^2 - 4\ln(x+1) + C .
\end{aligned}
$$

Also $f'(x) = 0 \Rightarrow x = 1$, which fixes $f(1) = 1$, and so $C = 4\ln 2 - 1$. Thus, we have

$$
f(x) = \frac{1}{2}(1+x)^2 - 4\ln(x+1) + 4\ln 2 - 1.
$$

(b)

$$
\begin{aligned}
f(x) &= \int 7 \sin 2x \cos 2x \, dx \\
&= \frac{7}{2} \int \sin 4x \, dx \\
&= -\frac{7}{8} \cos 4x + C .
\end{aligned}
$$

Also $f'(x) = 0 \Rightarrow x = 0$ or $\pi/4$ with $y = 1$. So we can fix $C = \frac{15}{8}$ or $C = \frac{1}{8}$ for the two cases respectively.

---

11. (a)

$$
\begin{aligned}
x(t) &= \int v(t) \, dt = \int (2 - t^2 + 3t) dt \\
&= 2t - \frac{t^3}{3} + \frac{3}{2}t^2 + C.
\end{aligned}
$$

When $t = 0$, we have $x = 2$ so we fix $C = 2$. The displacement is

$$
x(t) = 2 + 2t + \frac{3}{2}t^2 - \frac{t^3}{3}.
$$

(b) We have to solve $x(t) = 2+2t+\frac{3}{2}t^2-\frac{t^3}{3} = 2$ for t. This implies $-\frac{t}{6}\left(2t^2 - 9t - 12\right) = 0$, which gives the solutions $t = 0, \frac{9\pm\sqrt{177}}{4}$. The positive solution is $t_1 = \frac{9+\sqrt{177}}{4} \approx 5.576$.

(c) For $v(t) = 2 - t^2 + 3t = 0$,

$$
\begin{aligned}
0 &= 2 - t^2 + 3t \\
\Rightarrow t &= \frac{-3 \pm \sqrt{9 - 4(-1)(2)}}{-2} = \frac{3 \mp \sqrt{17}}{2} \\
\therefore t_2 &= \frac{3 + \sqrt{17}}{2} \approx 3.562 .
\end{aligned}
$$

(d) The plot is:

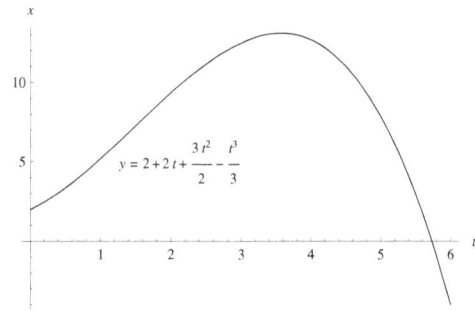

Figure 8.19: Plot of $x = 2 + 2t + \frac{3}{2}t^2 - \frac{t^3}{3}$.

(e) From part (d), we see that from t_2 to t_1 the displacement is always positive, so the distance moved is given by $D = |x(t_2) - x(t_1)| \approx 11.09$. We can also obtain the distance by integrating over the velocity vector, that is, $\left| \int_{t_1}^{t_2} (2 - t^2 + 3t) \, dt \right| \approx 11.09$.

---

13. Given $x(t) = 2\sin t$.

(a)

$$
\begin{aligned}
x(t_1 \to t_2) &= 2\sin(3\pi/2) - 2\sin 0 \\
&= -2 \text{ metres.}
\end{aligned}
$$

(b) Between times $t = 0$ and $t = \pi/2$ the particle is moving away from the origin. It comes to momentary rest, that is, $v(t) = 2\cos t = 0$ at $t = \pi/2$ and then moves towards the origin. The distance it travels is

$$
\begin{aligned}
D &= |x(t = 0 \to t = \pi/2) \\
&\quad + |x(t = \pi/2 \to t = 3\pi/2)| \\
&= 2 + |2(-1) - 2| = 6 \text{ metres.}
\end{aligned}
$$

(c) Displacement is a vector. It only depends on the initial and final positions of the particle. Distance is scalar quantity: It depends on the path.

---

15. Given $\frac{dx}{dt} = 2t \Rightarrow x = t^2 + C$. As $x = 1$ when $t = 0$, we can fix $C = 1$ and thus $x = t^2 + 1$. So, by the Chain Rule, we have

$$\frac{dy}{dt} = \frac{dy}{dx} \cdot \frac{dx}{dt} = \left(1 + \frac{1}{x+1}\right) \cdot (2t)$$

$$= \left(1 + \frac{1}{t^2 + 2}\right) \cdot (2t) .$$

$$\left. \frac{dy}{dt} \right|_{t=2} = \frac{14}{3}.$$

---

17. (a) For a straight line $y = mx$, we have $y'(x) = m$ and thus

$$L = \int_0^1 dx \sqrt{1 + m^2} = \sqrt{1 + m^2} \left[x\right]_0^1$$

$$= \sqrt{1 + m^2}.$$

Now, at $x = 1$, $y = m$. The distance between point $O(0,0)$ and $P(1,m)$ is given by $L = \sqrt{(1-0)^2 + (m-0)^2} = \sqrt{1 + m^2}$, which agrees with the above formula.

(b) For $y = x^{3/2} \rightarrow y' = 3\sqrt{x}/2$,

$$L = \int_0^1 dx \sqrt{1 + (3\sqrt{x}/2)^2}$$

$$= \int_0^1 \sqrt{1 + \frac{9x}{4}}$$

$$= \frac{8}{27} \left[\left(1 + \frac{9x}{4}\right)^{3/2}\right]_0^1$$

$$\approx 1.44 .$$

---

19. Direct differentiation of $y = xe^x$ gives $dy/dx = xe^x + e^x$. Thus,

$$\int_0^1 xe^x \, dx = \int_0^1 \frac{dy}{dx} - \int_0^1 e^x \, dx$$

$$= \left[xe^x\right]_0^1 - \left[e^x\right]_0^1 = 1.$$

21.

$$\frac{d}{dx} \ln(\sec x) = \frac{1}{\sec x} \cdot \sec x \tan x$$

$$= \tan x.$$

Thus we can evaluate the following easily as,

$$\int_0^{\frac{\pi}{4}} \tan x \, dx = \int_0^{\frac{\pi}{4}} \frac{d}{dx} \ln(\sec x) \, dx$$

$$= \left[\ln(\sec x)\right]_0^{\frac{\pi}{4}}$$

$$= \ln(\sec \frac{\pi}{4}) - \ln(\sec 0) = \ln \sqrt{2} \approx 0.35 .$$

Did You Know?

The integration symbol $\int$ arises from an elongated letter S.

The mathematician Leibniz, who invented calculus independently of Newton, chose that symbol to represent the fact that integration may be viewed as an infinite Sum, such as when one calculates the area under a curve by a sum of thin rectangles whose width is then taken to zero.

www.ingramcontent.com/pod-product-compliance
Lightning Source LLC
Chambersburg PA
CBHW051231200326
41519CB00025B/7333